WOMEN IN DAIRY DEVELOPMENT

ABOUT THE AUTHOR

Dr. (Miss) SHAGUFTA JAMAL is Research Associate in the Division of Extension Education, Indian Veterinary Research Institute, Izatnagar. After her B.Sc. (Home Science) from G.B. Pant University of Agriculture and Technology, Pantnagar, she worked as Teaching Associate at the same University. She then worked as Research Associate in the Ford Foundation Funded Project. She worked for several years as Compere (Announcer) in Farm and Home Unit of All India Radio, Rampur (U.P.).

Dr. Jamal has been the recipient of ICAR Fellowship during her B.Sc., M.Sc. and Ph.D. programmes. She has written many papers in research journals and technical articles in newspapers and magazines. She has presented papers in National and International Seminars and Conferences.

WOMEN IN DAIRY DEVELOPMENT

SHAGUFTA JAMAL
Research Associate
Indian Veterinary Research Institute
Izatnagar

Foreword by
Dr. P.N. BHAT
Deputy Director (Ani. Sc.) ICAR
&
Animal Husbandry Commissioner
Government of India
New Delhi

CONCEPT PUBLISHING COMPANY, NEW DELHI-110059

The Indian Council of Social Science Research gave the financial support towards the publication of this work. However, the responsibility for the contents of the book — facts, opinions, conclusions, etc. — is entirely that of the author and not of the ICSSR.

ICSSR Consultant : Dr. K. Sardamoni

ISBN 81-7022-500-0

First Published 1994

Published and Printed by
Ashok Kumar Mittal
Concept Publishing Company
A/15-16, Commercial Block, Mohan Garden,
NEW DELHI-110059

Phones : 25351460, 25351794
Fax : 091-11-25357103
E-mail : publishing@conceptpub.com

Dedicated to

My Brother

Syed Mazahir Ali

FOREWORD

Participation of women in agriculture in general and in animal husbandry in particular has always been appreciated but without any recognition and recording their contributions. By and large, they have remained "invisible workers". It is unfortunate that because of devaluation of their contributions, the rural women are denied their rightful status as active producers and access to development resources and services such as technology, credit and training. All this signifies the importance of having a fuller understanding of the role and contribution of farm women. The common view is that the women have a physical incapability rather than cultural limitation to interact with technology. The rural women, by and large, have, therefore, been excluded from educational and training programmes for rural transformation through the introduction of technology.

The major activities related to feeding and management of livestock are looked after by women. The technologies related to care of young calves, detection of oestrus in cows and buffaloes, symptoms of disease and management practices to reduce calf mortality, improving nutrition value of low quality roughages and other crop residues, conservation of fodder, disinfecting animals and animal sheds, clean and safe milking, processing of milk and preparation of milk products, etc. are relevant to farm women. Livestock enterprises with appropriate production technologies offer immense potential for socio-economic transformation for rural women.

It is often said by male agricultural professionals that the technologies have not reached rural women because women are not educated, they are not organised and they are not assertive. But, this is not true. Women in Andhra Pradesh dairy cooperatives – illiterate, poor with very few assets other than one or two milk cows or buffaloes – successfully, competently are carrying out artificial insemination, fat-testing and record-keeping. Rural women might be docile in male dominated organizations, but they are fully active, vocal and competent in running their own organizations.

The National Perspective Plan for Women (1988) describes the challenges ahead observing that "In case of both agriculture and animal husbandry, development strategies have provided very little attention to women in comparison to their active involvement in both the sectors". Empirical studies on the role of farm women are, therefore, required to produce adequate data for the planners and administrators to place proper emphasis on women specific enterprises and women specific technologies, thereby including the so far neglected 'women' in the transformation of the villages and in the overall development of the country.

Dr. (Miss) Shagufta Jamal, the author of this book has completed this unique study – on this subject disclosing the dynamic role played by farm women contributing to dairy productivity and dairy development. Dr. Jamal deserves my best compliments. I am glad that Indian Council of Social Science Research has taken up the task of publishing this work in the form of a book which could be a basis in the hands of planners and administrators to formulate strategies for dairy development in India. This book will be useful to the students, researchers and the extension professionals and agencies in developing scientific and realistic programmes.

DR. P.N. BHAT
Deputy Director General (Ani. Sc.) ICAR
&
Animal Husbandry Commissioner
Government of India
New Delhi

ACKNOWLEDGEMENTS

I express deep sense of gratitude to Dr. H.P.S. Arya, Head, Division of Extension Education, Indian Veterinary Research Institute, Izatnagar, for his worthy guidance, untiring help, valuable suggestions, keen interest and constant inspiration throughout the course of research work and in the preparation of the manuscript.

I am indebted to Dr. P.N. Bhat, Director and Vice-Chancellor, Indian Veterinary Research Institute, Izatnagar for his moral support and constant encouragement. Dr. Bhat, presently as Deputy Director General (Ani.Sc.), Indian Council of Agricultural Research, New Delhi, has kindly agreed to write the 'Foreword' of this book.

I am extremely grateful to my guide Dr. O.N. Kunzru, Joint Director (Extn. Education), Dr. Parmatma Singh, Head, Division of Livestock Economics and Statistics and Dr. Rajendra Singh, Scientist S-3, Division of Livestock Economics and Statistics as the members of my advisory committee for their valuable guidance and help in conducting the study and writing of this book.

I would like to record a special word of gratefulness to Dr. O.S. Verma, Sr. Scientist, Division of Extension Education, Indian Veterinary Research Institute, Izatnagar, for his valuable guidance and personal help during the course of study.

I am indebted to Dr. S.P. Singh, Associate Professor, Department of Agricultural Economics, G.B. Pant University of Agriculture and Technology, Pantnagar for the help and encouragement provided during the course of study. Special thanks are due to Dr. P.N. Kaul, Principal Scientist and Dr. M.K. Mandape, Scientist (S.G.), Division of Extension Education, Indian Veterinary Research Institute, Izatnagar for their cooperation and practical help provided from time to time.

Grateful thanks are due to Dr. I.H. Siddiqui, Sr. Scientist, Division of Biological Products, Indian Veterinary Research Institute, Izatnagar for his moral support. I owe my indebtedness to Mr. Yusuf Siddiqui, Sahabzada Khursheed Hussain and Mr. Farid Uddin Siddiqui for their help and support during the course of study.

I extend my heartiest thanks to Mr. Ajay Shukla, T-4, Division of

Extension Education for his statistical help and personal encouragement. I am also thankful to the staff and students of the Division of Extension Education, Indian Veterinary Research Institute, Izatnagar for their help and cooperation provided during the course of study.

No words can express the deep gratitude of my beloved mother, maternal uncle, Syed Abdul Qayyum Naqvi, Syed Absar Naqvi, brothers – Syed Nazahir Ali, Syed Mazahir Ali, Syed Mussawir Ali, Syed Ghazanfar Ali and Syed Moamber Ali, sister and *Bhabhies* for their affectionate support and encouragement during the course of the research work. They graciously boosted up my morale when I needed it the most to keep going towards my goal.

I also take this opportunity to extend special thanks to my uncle, Nawab Mamnoon Hussain, Ex-Honorary Magistrate, Aunts Begum Shakeela Hussain Begum Abdul Wajid, Ex-M.L.A. for their help and moral support during the course of my study.

Last but not the least I would like to record my sincere thanks to the respondents of the study for their cooperation in the collection of the data.

SHAGUFTA JAMAL

CONTENTS

LIST OF TABLES

LIST OF FIGURES

1

INTRODUCTION

"To awaken people, it is the women who must be awakened. Once she is on the move the family moves, the village moves, the Nation moves".

— *Jawaharlal Nehru*

Historians believe that it were women who first started growing crops and domesticating animals and thereby initiated the art and science of farming (Swaminathan, 1985), while men went out for hunting in search of food, women started playing with animals, gathering seeds from the native flora and began keeping the animals and cultivating those interests from the point of view of foods, fodder, fibre and fuel. Women have played, and continue to play, a key role in the conservation of basic life support systems such as land, water, food and shelter. That women play a significant and crucial role in agricultural development and allied fields, is a fact long taken for granted but also long ignored. Studies on women in agriculture conducted in India and other developing countries have indicated that women contribute far more to agricultural production than has generally been recognised or acknowledged. By and large they have remained "invisible workers". Because of low "visibility" of their contributions, they are not regarded as equal partners in the developmental processes. It was in this context that the United Nations had declared the ten-year period ending 1985 as the "Decade for Women" with a view to "promote genuine equality between men and women". Since then the women have been engaging some attention, at all levels, for their role in rural development. Even today, as in the past, the household activities of Indian women, in general, and farm women, in particular, absorb a very large part of their time and energy. Certain economic activities particularly related to agriculture, such as caring for domestic animals, maintaining kitchen garden, collecting manure, fetching fuel and water, etc., are also considered under household activities.

If we look into the history, we find that men have usually had first choice of jobs, certain jobs have remained closed to women simply because men preferred these jobs. In male dominated society he succeeded in reserving the same for themselves. It is for this reason that women are found doing jobs which are hazardous, monotonous and repetitive in nature and carry low prestige. Such tedious and laborious jobs as transplanting, weeding, harvesting, cleaning and processing are mostly done by women. Certain cultural factors like belief that women are delicate and should be assigned easier and lighter tasks, contributed to the determination of the societal role of women. At times, it so happened that when some traditional women's jobs, such as harvesting, processing etc., are mechanised and consequently involve less drudgery, they are taken over by men.

There is no denying the fact that women have suffered utter neglect and deprivation over centuries. The passive role played by women in the male dominated society has resulted in the present state of general neglect of women. Swaminathan (1985) observed that "despite their importance in agricultural production women face severe hardships. They are the largest group of landless labourers with little real security in case of break-up of the family through death or divorce, inheritance laws and customs discriminate against them. Land reforms and settlement programmes give sole title, and hence the security, to males. Extension workers, almost exclusively male, aim their advice at men and men's activity". Some positive thinking by the Government has, however, recently gone into the inadequacies in the inheritance laws, property rights and representation on civic bodies with respect to women.

Bhatt (1988) observed that farm women, with their land based, livestock based and home based occupation opportunities, are in a position to generate additional income and employment for the family. Modernization and rapid technological changes have affected women's role adversely. The introduction of new technology requires new skills. Women do not have access to training in these skills. Hardly any research studies have been undertaken to evolve suitable technology packs which will improve the employment conditions and efficiency of women, alleviating drudgery and/or generate employment opportunities so that they can add to the family income either directly or through increased land/livestock productivity.

Women participate, at different levels, in almost all activities of crop husbandry right from preparatory tillage to harvest and even in post-harvest tasks such as processing, storage and marketing of the agricultural produce, the only exception perhaps being ploughing. Operations such as seed cleaning, seed grading, sowing, transplanting, weeding, gap filling, irrigating, harvesting, threshing, shelling, hulling, winnowing, feeding cattle, harvesting and collecting fodder and grasses, grazing the animals, looking after milch animals, providing them drinking water, cleaning of sheds, milking, processing, storage and marketing of milk and milk products, etc. are all primarily shared by farm women.

Nearly half of the available global human resource is women. According to 1981 Census, the female population in India was 333.79 million (47.23%) of which 253.72 million (77%) belonged to rural areas. In India 222.52 million (33.45%) were main workers and 22.09 million (3.22%) marginal workers. Among the main workers, 177.55 million (79.80%) were males and 44.97 million (22.20%) females, and of the marginal workers, 3.54 million (15.57%) were males and 18.55 million (84.43%) females. About 81.23 per cent of female and 65.60 per cent of male main workers were engaged in agricultural sector; and 18.77 per cent of female and 34.40 per cent of male main workers were engaged in non-agricultural sector. The reliability of the data pertaining to actual women workers, is always questionable in Indian rural society where their values and/or prestige does not permit them to commit that the women members in their family work on farm or even contribute in livestock production activities. The women in India, like in many developing countries, are silent workers. Low literacy rate among rural women and their traditional cultural bondages further confound the generation of data. Hence, the actual contribution of rural women in farm activities is further belied by the available statistics. The percentage of female literates in rural areas, according to 1981 Census, was 18 per cent as against 40 per cent for males in rural areas. More than 75 per cent young rural women belong to the families of small and marginal farmers and landless labourers and more than two-thirds population of rural women is illiterate. Illiteracy, ignorance and poverty are often responsible for keeping vast number of Indian rural families in the State of underdevelopment and consequently vulnerable to poor health.

It is difficult to exactly quantify the role of farm women because of widely varying agro-climatic conditions, cropping seasons, socio-economic and cultural differences and changes in concepts and definitions used in the official census and labour survey reports. However, the serious implications of their invisibility for their growth and development, as also for effective utilization of their potential, provides the most compulsive logic for improving statistics on their role.

Livestock production, is equally important to both land-owning and landless families. Milk and milk products play a vital role in the country's agricultural economy. Dairy provides sustenance to millions of farmers, particularly the rural poor. Dairy development fits most appropriately in the country's programmes of increasing rural employment and equitable distribution of income with justice. In small farm situations also milk and milk products directly help in increasing crop production through making fluid cash available to the farmer for the purchase of essential inputs.

India's bovine population, according to 1982 Census, was 191 million cattle and 69 million buffaloes out of which, the milch cows and buffaloes were 50.71 and 28.32 million, respectively. India, with an annual milk production of 43.90 million tonnes in 1986-87, ranks third in the world after Soviet Union and the United States. But, in India, the average annual milk yield per cow is 157 kg and that of buffalo is 504 kg as against 800 kg for non-descript cows, 1800 to 2400 kg for cross-bred cows and 1250 kg for buffaloes in the organised sector where dairy animals are reared for commercial milk production (NCA Report, 1976, p. 67). The productivity of our rural milch animals is, in general, very poor. Increasing the productivity of these animals is, therefore, the greatest challenge before the scientists and the researchers.

It is unfortunate that because of distorted and/or partial information about their involvement in agriculture and consequent devaluation of their contribution, the rural women are denied their rightful status as active producers in agriculture and access to development resources and services like technology, credit and training, further contributing to their marginalization. All this signifies the importance of having a fuller understanding of the role and contribution of farm women so that the extension services may accordingly be planned to fully integrate them in the development process by better serving their specific needs, interests and capabilities. The women have so far been erroneously viewed as passive or neutral functionaries in the socio-economic and technological transformation processes and programmes being planned and implemented by planners, administrators, economists, funding agencies and extension agencies. Their role in all kinds of rural economic activity has been, to say the least, ignored. The common view is that the women have a physical incapability rather than cultural limitation to positively interact with technology. The rural women, by and large, have, therefore, been excluded from educational and training programmes for rural transformation through the introduction of technology.

It was the Sixth Five Year Plan document in which some attention, for the first time, was directed to the untapped women resource and a separate chapter on women was devoted to it. In the Seventh Five Year Plan its importance has further been recognised. Special attention has been given to provide them technical skills and training in various fields in order to technically equip them to actively participate in and take up economic activities under their socio-economic situation. The National Perspective Plan for Women (1988) describes the challenges ahead in the following words: "Agriculture and allied fields provide the largest sector for women's employment. It largely determines the rural women's socio-economic status. In case of both agriculture and animal husbandry, development strategies have provided very little attention to women in comparison to their active involvement in both the sectors".

This underlines the need to empirically know the role of farm women in animal husbandry sector which has become even more complicated to understand in view of rapid changes in the socio-economic milieu in view of the available technology.

1.1 Statement of the Problem

The area of cattle and buffalo production has traditionally been a responsibility predominantly of farm women. In spite of their major role in cattle and buffalo production they have been left out from extension programmes related to animal husbandry. This is one of the major reasons for making very slow progress in this sector. The planners and administrators did not give proper recognition to the rural women for their role in dairy cattle and buffalo raising due to lack of realistic data and information on it and the factors governing it. In today's fast changing society their role needs to be empirically studied in its total dynamic dimensions. Role of any individual in a system is dependent upon several factors including one's own perception, expectations and past performance. The role of house wife, being the partner of her husband, depends upon the expectations of her spouse.

Keeping the above in mind, this study was taken up through which efforts were made to critically study the role of farm women in relation to productivity of dairy animals, in a given social system, wherein various operating elements, including the perceptions and expectations of the roles of related individuals have their dynamic effects on the actual performance of those roles.

1.2 **Objectives of the Study**

General

The broad objective of the study was to provide useful and empirical information about the role of farm women in relation to productivity of milk by dairy animals in rural social system.

Specific

Based on the above general objective, the following specific objectives were framed for the study:

1. To study the role expectations of spouses from farm women in relation to dairy animal production tasks.
2. To study the role of farm women as perceived by themselves in relation to dairy animal production tasks.
3. To study the extent of involvement of farm women and other family members in decision-making in relation to dairy animal production tasks.
4. To find out the levels of knowledge and skill of farm women about selected dairy husbandry practices.
5. To study the role performance of farm women in relation to productivity of dairy animals.
6. To study the dynamic interplay of various factors influencing the role of farm women in relation to productivity of dairy animals.
7. To identify the training needs for farm women in relation to dairy animal production.

1.3 **Practical Utility**

The present study on role dynamics of farm women in relation to productivity of dairy animals will yield useful empirical information on role relationship between farm women and their spouses in relation to cattle and buffalo production under rural situations. Further, it is expected to yield information about what farm woman perceives herself, what was expected of a rural wife by her husband and how far she was performing according to the expected pattern with respect to productivity of cattle and buffaloes. This information will be useful in assessing the attitude of rural livestock owners towards rural women and their role patterns in cattle and buffalo production.

A cognitive picture of farm women could be built by finding the role played by farm women and their spouses in dairy animal production. Identifications of the levels of knowledge and skills required to be applied in respect of their respective involvement in decision-making process would help in developing realistic developmental and training programmes for them.

The study will help in formulating plans for improving performance and production of milch animals and hence in generating income and additional employment for farm families.

The study will be of immense use to home scientists, educationists, extension personnel and research workers as well as planners and administrators who are involved directly or indirectly in attempting to stimulate and catalyse the adoption of improved technologies and practices with respect to cattle and buffalo production in rural areas.

1.4 Limitations

(i) This study has the usual limitations of time and resources to be encountered by the student researcher.

(ii) The items included in the study for detailed investigation are also limited since it is not possible to cover all the areas in such a short time.

(iii) The findings of the study can be generalised to the area of study from which the random sample has been drawn and other areas with similar social, cultural and economic systems in operation.

2

REVIEW OF LITERATURE

The subject of role of farm women in dairy animal production has wide potentialities for scientific enquiry. No study has been undertaken in a systematic way in this direction. An attempt has, however, been made in this chapter to review the studies relevant to the present study.

For convenience, the review of literature has been presented under the sub-heads viz. role of farm women in agriculture and dairy animal production and their characteristics, economic motivation, information source utilization, decision-making pattern, role expectations, role perception, knowledge and skill of farm women, productivity of milk and role performance.

2.1 Role of Farm Women and their Socio-personal Characteristics

Farm women are engaged in considerable amount of activities of farming and allied occupation. Most of the studies conducted on this aspect highlight the fact that animal husbandry activities are more liked and performed by farm women than crop farming activities. In one of the study, the farm women were found to like churning and milking activities the most, and harvesting of crop as the least (Nainie, *et al.* 1966).

Women were reported to derive considerable opportunities for employment and additional income from animal husbandry enterprise in Kaira District of Gujarat for the farm family (Bhat, 1975). Income from two buffaloes was able to look after a landless widow and her family.

Studies conducted in respect of time devoted by farm women on animal husbandry activities have indicated that they devote a considerable time on these activities. Verma (1984) found that 3.16 hours per day time was devoted on animal husbandry activities by farm women. On the other hand Sridevi (1984) reported that they spent

32.92 per cent and 2.50 per cent of their time on farm and allied activities including animal husbandry. Veena *et al.* (1986) found that the rural women spent 7.5, 4.0 and 6.7 hours per day in home, dairy and allied activities respectively. In peak season the average time spent by them on these activities increased to 6.4, 3.6 and 4.00 hours per day respectively. Lepcha (1987) reported that average daily hours spent on agricultural and livestock production activities were about 7.20 hours per day.

A study conducted in Bareilly district of U.P. concluded that on an average, farm women spent 3 hours 3 minutes per day on animal husbandry activities during lean agricultural season while during peak agricultural season they spent 2 hours 53 minutes per day on these activities (Agrawal, 1988). In another study, the total time devoted by farm women on agriculture, animal husbandry and supportive activities was found to be 17 per cent, 68 per cent and 15 per cent respectively (Arya and Madhukar, 1988). Patel and Mehta (1988) reported that landless women work more number of hours in dairying than men and their per hour earning from dairying was four times more than from agriculture. They spent 40 per cent of their time in dairying enterprises whereas women from small farmer category spent 33 per cent of their time.

Studies have indicated that women participation in agriculture and animal husbandry activities depends on various socio-economic factors of the respective families. Rao (1976) reported that rural women's participation was negatively correlated with the size of their family and land-holding. Malathi (1979) also found relationship between women participation and caste, age, education and extension contact. Women's socio-economic status was found to be correlated with the adoption of improved practices (Bhangoo, 1984).

The socio-economic and personal characteristics were found to vary in women participating in different activities. The maximum number of women participating in storage, harvesting and threshing belonged to age group of 25-40 years and family size of 5-8 members. Number of rural women participating in any of the farm activities was more in nuclear families than joint families. The maximum number of rural women participating in storage, plant protection and sowing belonged to backward class, while maximum number of rural women engaged in threshing, harvesting and irrigation belonged to scheduled castes (Sirohi, 1985).

Livestock holding, family educational status, economic status and risk orientation were reported to have positive and significant correlation with the adoption of H.S. vaccination by livestock owners (Kunzru *et al.*, 1987). Total work load on women decreased with the increase in the size of the holding. On the other hand men's total load increased with the increase in the size of holdings. The difference was reported to a large extent in agricultural work load and to a smaller extent in carrying for cattle (Saraswathi, *et al.*, 1987).

Involvement of women from scheduled caste families in livestock rearing was found prominent (Dasgupta and Sengupta, 1988). Their level of education and exposure to extension were also found low. Similarly, Kanwar and Guleria (1988) also found that the number of female workers per farm per hectare was highest in case of scheduled castes. Illiteracy was also higher in scheduled castes.

2.2 Economic Motivation

Economic motivation of the farmers have been found to have positive and direct effect on the adoption of improved practices. Kolte (1967) observed that a large number of farmers who desired to earn more profit adopted more improved farm practices. Similarly, high economic motivation was found responsible for high adoption of new dwarf varieties of wheat (Roy, 1972).

Farmers used to adopt improved practices for economic gain (Das and Sarkar, 1979). In dairy, fodder and cereal crop production, 69 per cent farmers had high level of economic motivation (Fulzele, 1986).

2.3 Information Sources Utilization

Information input is directly related to adoption of improved practices. Study conducted by Kunzru, *et al.* (1987) reported that information source utilization was positively and significantly correlated with the adoption of H.S. vaccination by livestock owners. But, unfortunately information input for women is very low. Bhuiyan (1988) reported that access to information, improved technology and credit facilities and participation in extension related activities appears to be almost nil among rural women.

Majority of farm women of small and marginal farmer families were reported to get information only through their friends, relatives and neighbours and did not get information from institutional sources (Jamal, 1984). Similarly, the most used sources of information by small farmers were family friends and progressive farmers followed by neighbours, demonstration, V.L.Ws. and radio (Singh, 1971). A study on media utilization pattern of the farmers conducted by Mathur (1974) reported that there was a predominant use of interpersonal source of communication such as neighbours, friends and relatives, block and IARI personnel in the process of decision-making. It was further revealed that farmers with lower middle socio-economic status used more of neighbours and friends as information sources especially at the initial stages of decision-making. It was concluded that less literate farmer used neighbours, friends and relatives more as source of information. Mass media had been mostly used by the farmers having primary level of education and above.

Non-institutionalized sources of information were reported to rate high over institutional sources of information by the farmers in the advanced stages of adoption (Jha and Singh, 1965). Personal cosmopolite sources were found to be used to a greater extent by large farmers than small farmers (Sawhney, 1967). In another study conducted by Singh (1970), similar results were observed. Personal cosmopolite sources provided information to majority of people in progressive villages whereas in non-progressive villages these sources provided information to only few people.

Small farmers used to have poor information input which is deterimental for increasing agricultural production (Singh and Ambastha, 1975). Education, socio-economic status, farm size, risk preference and cosmopolite-localiteness were found to be significantly correlated with information input.

A few studies have reported field level workers as an important source of information for rural men and women. Barua (1982) reported that Hindu farm women contacted V.L.Ws., traders and money-lenders whereas Muslim farm women contacted traders, input dealers and VLWs. Upper Hindu women contacted all the sources frequently as compared to lower Hindu and Muslim women. A pilot study on the role and awareness of farm women conducted in Jammu Division (Anonymous, 1988) observed that the field functionaries of the agriculture department were helpful to the farmers as they provide them advice, seeds and fertilizers.

2.4 Decision-making Pattern

In the male dominated society normally the decision is taken by men only. It has however been reported that while taking the decision the women are consulted in various activities such as seed storage, care of animals (Sharma and Singh, 1970) marketing and storage of farm produce, sale and purchase of animals, purchase of fertilizers, sowing time and method of sowing and intercultural operations (Nand, *et al.*, 1988).

Younger male family heads depended more on female family members for decision-making than old family heads. Consultancy of women decreased with the increase in the age of male family heads. However, the women continued to influence the decision regarding home management irrespective of the age of male family heads (Arya, 1964). Older women were always consulted more than young and middle aged women. Only one third of women were found to be consulted sometimes (Gupta, 1986).

Wives were reported to have an appreciable influence in decision-making in the lower strata and nuclear families (Dey, 1977). Rural women were found to have varying degrees of participation in decision-making in various activities (Dubey, *et al.*, 1982). Their participation in decision-making was high in areas of utilization of milk, number of milch animals to be kept, quantity of milk to be sold and kept for home consumption and type of fodder to be fed. In areas of fixing the rate of milk, feeding of animals at calving time, which animals to be sold, feeding of young calves, quantity of fodders and concentrates to be fed to milch animals, breed of milch animal to be purchased, care of new born calf, source of drinking water for animals, stopping of milking before calving, number of times of milking, treatment of animals and housing of animals in cold and hot seasons, the participation of rural women was medium. Their low participation in decision-making included the areas of type of housing for animals, utilization of income from milk, vaccination against contagious diseases, grazing of animals and breeding practices in animals.

Participation of women in decision-making was not significantly associated with caste and family income. Their participation was higher in nuclear families and in lower socio-economic status families (Yadav, 1982). Self esteem influenced the extent of participation in decision-making in respect of farm animals and farm machinery (Gulabani, 1987).

Ghosh and Adhikary (1988) reported that in general women are still getting least importance in decision-making process excepting borrowing money, seed storage and some of the practices related to livestock like determining the size of livestock holding, type of livestock to be raised, selling of livestock, and care and weaning of calves/kids etc.

Farm women's involvement in decision-making process was the highest in purchase and sale of animals followed by storage of farm produce (Kaur, 1983). None of the farm women were involved in decision-making regarding variety of crop to be sown and farm credit.

Farm families specially the low and middle income families were reported to take the decisions jointly (Cravon, 1963). All members of farm families took part in decision-making in the agricultural practices related to production and storage, sale of produce and purchase and sale of livestock (Sunita, 1977). In another study it was concluded that in majority of the cases relating to home, as well as farm, husband and wife jointly took decisions (Badiger, 1979). Seema (1986) also found that the joint decisions were made regarding purchase and sale of land, care and management of animals and children education. In practices related to livestock, male heads of families were reported to take the decision jointly with their wives (Ghosh, 1987). In the areas of animal health most of the decisions were taken jointly (Agarwal, 1987). Husband and wife together used to decide the issues concerning seed storage, care of animals, selection of seed and time of selling farm produce (Singh, 1968). Joint decisions by husband and wife were also reported to be taken in respect of artificial insemination, feeding of balanced ration, exercise of animals, disinfection of sheds and deworming of animals (Verma, 1984).

Comparatively larger proportion of women participated in decision-making with respect to seed storage, winnowing, care of animals and harvesting (Singh, 1968). Chole (1973) found that keeping of milch animals, employment of daily paid labour, processing and harvesting of agricultural produce, selection of different crops for cultivation were, in general, the important issues relating to farm in which all categories of farm women and particularly mothers played important role in decision-making.

The decisions regarding the number of cattle to be kept (97%) and selling of milk and milk products (89%) were reported to be made mostly by wives while the decisions on treating sick animals and buying and selling of animals were predominantly taken by husbands (Puri, 1972). Women were found generally as decision-makers in the subsidiary enterprises like maintenance of livestock while on the main enterprise, i.e. agriculture the decisions were made by male members of the family. A majority of farm women wished to take decisions alongwith their husbands or other menfolk (Jhansi Rani, 1981). Saraswathi, *et al.* (1987) found that the decisions regarding the home consumption or selling of livestock produce were made by women. She further observed that sale proceeds were generally kept with the women and spent by them on what they considered as appropriate.

Mothers-in-law were expected to do more decision-making tasks alone than the daughters-in-law. On the other hand, the daughters-in-law were expected to do more of physical tasks, both home as well as farm related and less of decision-making tasks. Mothers-in-law perceived the daughters-in-law performing the decision-making tasks significantly more than expected of her. But, the daughters-in-law perceived the mothers-in-law performing the decision-making tasks at the level expected of her (Garg, 1966). Chand (1967) also found that the mothers-in-law were expected to perform more decision-making tasks than the daughters-in-law.

2.5 Role Expectations

Role senders have expectations regarding the way in which the focal role should be performed. They also have perceptions regarding the way in which the focal person is actually performing and they correlate the two and exert pressure to make his performance congruent with their expectations (Kahn, 1960). In a comparative analysis of the role performance with the role expectations, performance fell short of the role expectations in all the four areas studied, viz., planning, supply and service, farm business and educational function (Sharma, 1971). In each case the difference between role expectation and role performance was significant at one per cent level of probability. Further, Bhatnagar (1971) observed that there were significant differences in the consensus among the role set on role expectations and role performance on majority of items.

Daughters-in-law were expected to perform more physical tasks whereas mothers-in-law and fathers-in-law were expected to perform more decision-making tasks. The views of mothers-in-law were found to be significantly similar on their expected roles in the study conducted by Garg (1966).

Rural women were found to be expected to perform eight major roles in farm and home management, namely, harvest and post-harvest activities, pre-sowing and sowing tasks, food preparation, child and family care, allied agriculture (poultry, dairy, kitchen gardening, etc.)

activities, intercultivation, house-keeping and social and leisure time activities (Devi, 1983).

2.6 Role Perception

The performance of an individual is affected by his own perception of role responsibilities. Kherde (1971) reported that more was the self perception of responsibilities by a VLW, higher was his supervisory role performance.

The village extension officers working with intensive agricultural extension programme (T&V) perceiving their job as of high social status had high achievement motivation and perceived their work load as average (Reddy, 1983).

Rural women from low economic category were found to perceive farm activities as an important activity for them whereas women from medium and high economic categories considered the allied agricultural activities as an important activity for them (Devi and Reddy, 1984). It was also observed that majority of rural women who belonged to the high and medium economic categories, did not go out to attend farm work. Status consciousness was operating on different strata of society, preventing them from attending farm work.

Studies have reported significant differences in the perception of role expectations among rural women, their husbands and experts (Devi, 1983).

2.7 Knowledge

Literates have better response than illiterates in improving their knowledge (Deshpande, 1962). Knowledge level of farmers about package of practices of wheat was reported to have significant association with their educational level. On the other hand no association was found between socio-economic status of farmers and their knowledge level (Singh, 1971). High level of knowledge was found in majority of farmers with high educational level, high family educational status, large land-holding, high exposure to mass media and high favourable attitude towards the respective enterprise, i.e. dairying (Sidhu, 1980). The majority of the respondents of old and middle age groups, with large family size, medium herd size and high extension contacts had medium level of knowledge. Age and family size were not associated with the level of knowledge.

Increase in knowledge about dairy innovations leads to higher adoption of dairy technology by dairy farmers (Tyagi and Sohal, 1984). Knowledge was found positively and significantly correlated with the adoption of H.S. vaccination by livestock owners (Kunzru, *et al.*, 1987). In earlier study also Sidhu (1980) reported that knowledge had positive and highly significant association with milk production per household per day, per household per year and per animal per day.

Most of the rural cross-bred cattle owners were not aware of scientific breeding, feeding and health practices as reported by Dubey and Singh (1975). They further reported that the merits of the cross-bred cows had not been properly propagated by the implementators of the ICDP programme in Karnal district. Most of the rural owners of the cross-bred cows were rather ignorant of scientific dairy practices. Their knowledge in breeding, feeding and health care was found to be very poor (Dubey and Singh, 1976). Gill and Singh (1977) in their study also reported low level of knowledge among dairy farmers regarding breeding, feeding, housing and animal health practices. Medium level of knowledge was however reported in management and marketing of milk among dairy farmers. Similar results were reported by Sohal and Tyagi (1978) while studying ICDP, Karnal and by Gite (1980) while studying training needs of dairy farmers. Knowledge was extremely poor in the labour group. A wide gap existed between potential knowledge and the knowledge acquired by the cattle owners (Sohal and Tyagi, 1978).

About 97 per cent of the farmers were found to have low level of knowledge about feeding practices (Rath, 1978). Further, 88 and 12 per cent of the farmers were found to possess medium and high level of knowledge respectively about breeding practices. On the other hand Nataraju and Channegowda (1985) reported that more than 95 per cent farmers had known about A.I. in cattle. More than 80 per cent of cattle owners had known about improved milch breeds, disease occurrence and precautionary measures to be taken against contagious diseases. Further, relatively larger percentage of agricultural labourers as compared to small and marginal farmers had no knowledge about all the above-mentioned animal health and production practices. Respondents were aware about the important practices like deworming of calves (66%), dehorning of young calves (54.70%), feeding of dry fodder (52%), feeding of green fodder (22.70%), improved fodder varieties (37.30%), cattle insurance (23.30%) and maintenance of important dairy management records (14%).

Based on the average knowledge score of milk producers, breeding, feeding, health care, management and fodder production were observed to be ranked in that order (Panwar, 1979). In a study of health cover practices of poultry, 26 per cent of the farm women possessed low level of knowledge while four per cent possessed medium level and none possessed high level of knowledge. Knowledge-age and knowledge-communication relationships were found to be positive but higher relationship was found between knowledge and age than between knowledge and communication (Rehman, *et al.*, 1988). Women generally are the carrier of traditional knowledge. Much of the rich store of knowledge about traditional eco-systems resides in women (Ordonez and Goldey, 1985).

2.8 **Skills**

Prospective agriculture needs skilled manpower. Education and training of women in agriculture is, therefore, imperative for meeting the future challenges (Singh and Punia, 1988). Medium level of skills were developed in the selected aspects of dairy farming immediately after the training programme among majority of trainees (Fulzele, 1986).

Literates have shown better response than illiterates in improving their skills (Deshpande, 1962). Necessary skills can be imparted to the selected traditional leaders (Dubey, 1977). Periodic training is extremely important. They may be provided necessary guidance and counselling in their day to day work. They should be looked upon as important links between the community and the trained professionals. Bhatnagar (1971) reported that the frequency of performance of SMS is related to skill, time spent and training needs.

2.9 **Productivity of Milk**

About 66 per cent of the total milk production in the country comes from buffaloes while there are only seven recognized breeds, covering only 15 per cent of world buffalo population (Bhatnagar, 1985). Productive ability of the cross-bred cows in rural areas was found unsatisfactory. Most of the cows had their peak yield between 5 to 11 kg per day. In nearly one-third of the cases they could not conceive in time (Sharma, 1983). The productivity of our rural milch animals was reported to be very poor by Dhillon (1982) also. For attaining high and economic levels of milk production and its further improvement,

Gurnami (1985) suggested that the farmers should get rid of their surplus and uneconomical animals.

Gill and Batra (1967) attributed low productivity of animals to insufficient feeding. Low intake of nutrients in the face of extremely limited supply of feeds was observed as the first constraint by Jackson (1978) contributing to low productivity of dairy animals. Dhillon (1982) however, observed that the low productivity is partly due to poor management practices followed in rural areas.

Finance, cost of inputs such as medicines, concentrates and mineral mixture, genetic potential of milch animals, efficiency of artificial insemination service, technical knowledge of animal husbandry practices for rearing the animals were perceived as the most important factors affecting productivity of dairy animals by all categories of milk producers, viz. landless, marginal, small and medium and large farmers (Sagar, *et al.*, 1988). Irrigation facilities for growing nutritious fodder crops, supply of good quality seed of fodder crops and feeding of nutritious green fodder, concentrated and mineral mixtures to milch animals were also perceived important factors affecting productivity by the marginal, small and medium-large categories of milk producers.

Productivity of milk of dairy cows was found to be positively and significantly correlated with the livestock holding (milch cattle) of the landless (Sagar, *et al.*, 1988) they further reported that the productivity of milk of dairy cows was positively and significantly correlated with the number of members in the family of the landless livestock owners. Family educational status had no significant relationship with productivity of milk of dairy cows. Productivity of milk of dairy cows was also found positively and significantly correlated with the economic motivation of the livestock owners and the information sources utilization by them.

Average income from milk and milk products, total annual income and feeding of the animals were reported to have positive and significant correlation with productivity of milk of dairy cows (Sagar, *et al.*, 1989).

Adoption of improved practices observed several constraints both at organizational as well as livestock owners level. Artificial insemination had many problems for its successful implementation in rural areas (Tomer, 1978). It was concluded that not more than 30 per cent of cows and 20 per cent of buffaloes were artificially inseminated. Ignorance of people about the role of genetic potential of animal in its productivity was reported to be the major problem in adoption of artificial insemination.

In another study, post-vaccination reduction in milk yield was found to be one of the important constraints in the adoption of HS vaccination as perceived by 81.60 per cent of the respondents (Kunzru, *et al.*, 1987).

2.10 **Role Performance**

In one of the study, men were found to perform majority of tasks that were associated with breeding, fodder cultivation, health and disease and marketing while the women did most of the tasks that were associated with feeding and management (Dey, 1977).

Rural women belonging to low economic category performed more work in farm management whereas those belonging to medium and high economic categories performed more work in house management (Devi, 1983). It was further found that caste and education were negatively and significantly associated and family size was positively and significantly associated with the farm and house management role performance of rural women. Bose (1985), however, found that though the female labour participation was influenced by caste factor but it was not the role governing factor in female employment outside home.

Participation in farm as well as house related physical activities was reported more for daughter-in-law as compared to mother-in-law and daughters of the household (Kaur, 1981). With the development of agriculture, the extent of participation of rural women in various domestic activities had decreased as now they were in a position to own various time-saving household convenience (Kaur, 1986). With regard to their participation in farm activities, there did not seem to be any noticeable change except that for the management of cattle their role was becoming more of supervisory nature as for these tasks they were employing hired labour. Satpathy (1987) reported that participation of farm women, especially having small holdings, was quite substantial in respect of all animal related activities. Involvement of women in milking and marketing was independent of land-holding.

Majority of the farm women covered in the study of Promila (1981) were found to participate in dairy enterprises and considered it profitable. Majority of them were giving green fodder and concentrates but not minerals.

As high as 83 per cent of women in the labour force were reported to participate in almost all agricultural operations including animal husbandry, dairying and allied activities and contribute to family and national economy (Jain and Shukla, 1988). Their contribution is unfortunately not recognized. It was pointed out that the participation of women in enhanced milk production will have little effect on dairy development if it is not coupled with proper processing, utilization and marketing of milk. The development and adoption of modern processing technologies tailored to preserve milk and conserve milk solids in rural areas and techniques for value addition and improvement in storageability and marketability of milk and milk products will guarantee the employment to farm women all the year round and protect their employment in the milk trade where it is declining due to technological advancements.

Share of female labour in total family labour was found to be highest in fodder collection activities while the share of male labour was higher in tending of farm cattle activities (Sharma and Muthuria, 1985). Majority of farm women have been playing a dominant role in about half of the 17 agricultural production related tasks studied by Dak, *et al.*, (1986). In order of their contribution these tasks were: tending farm cattle, collecting fodder for the cattle, selling livestock products and storage of produce, treatment of sick cattle, harvesting of crops and making farm yard manure. A majority of the respondents were playing monopolising role in the first three of these tasks. Gupta, *et al.* (1986) found that contribution of women in livestock rearing was in the areas of cutting, collection and chaffing of fodder, feeding the animals, cleaning the shed, milking, preparation of milk products, collection of cowdung and preparation of cowdung cakes and their storage. They also reported that cattle care was more or less a female domain as women played a key role in performing various cattle related tasks.

Women participation in farm and home activities was found maximum when their children were in age group of 5 to 10 years (Veena, *et al.*, 1986). They found that 60 per cent, 90 per cent and 95 per cent of women in farm families participated respectively in crop, dairy and home activities. Sardana (1988) observed that Indian farm women are playing dominant role in about eight agricultural activities. In descending order these are tending dairy cattle, collecting fodder from the field, selling livestock produce, storage of farm produce, weeding operations, management of cattle treatment, harvesting the crops and making of farmyard manure. Significantly, livestock related activities were more pronounced and the farm women had virtually monopolized the cattle care operations to the extent of almost 95 per cent. Similarly, another study conducted by Kumari, *et al.*, (1988) observed that 80 per cent of all important dairying operations in dairy milk production were performed by women. They also mentioned that traditionally, women sold milk to milk traders and middlemen and kept the milk money for their expenses.

Involvement of farm women was widespread and larger in respect of livestock related activities than crop related activities (Govind, 1984). Their involvement in all livestock activities was found to be more as compared to their involvement in crop husbandry practices (Govind and Subramanyam, 1988).

3

RESEARCH METHODOLOGY

This chapter deals with the sampling design, theoretical orientation of the research problem, the independent and dependent variables and their empirical measurements, tools of data collection and the statistical methods used in the analysis of data.

3.1 **Sampling Procedure**

3.1.1 *Locale of the Study*

Bareilly district in the state of Uttar Pradesh was selected purposively on the basis of following criteria:

i. In this region dairy animals are owned by majority of the rural population;
ii. Women play an important role in animal husbandry enterprises, especially with respect to dairy animals;
iii. Since the researcher stays in the district, she was familiar with the local language. It helps to build quick rapport which is necessary for data collection; and
iv. This district is well served by roads and passenger bus services.

3.1.2 *Background information about Bareilly district*

Bareilly district, in the Rohilkhand Division, is situated 237 km away from Lucknow, the State Capital. The district is situated between latitudes 28.10°N and 28.54°N and Longitude 78.58°E and 79.47°E. It is surrounded by Pilibhit district on East, Rampur district on West, Nainital district on the North and Shahjahanpur district on South side. The district is spread over about 150 km, from North to South and about 90 km from East to West (Fig. 3.1).

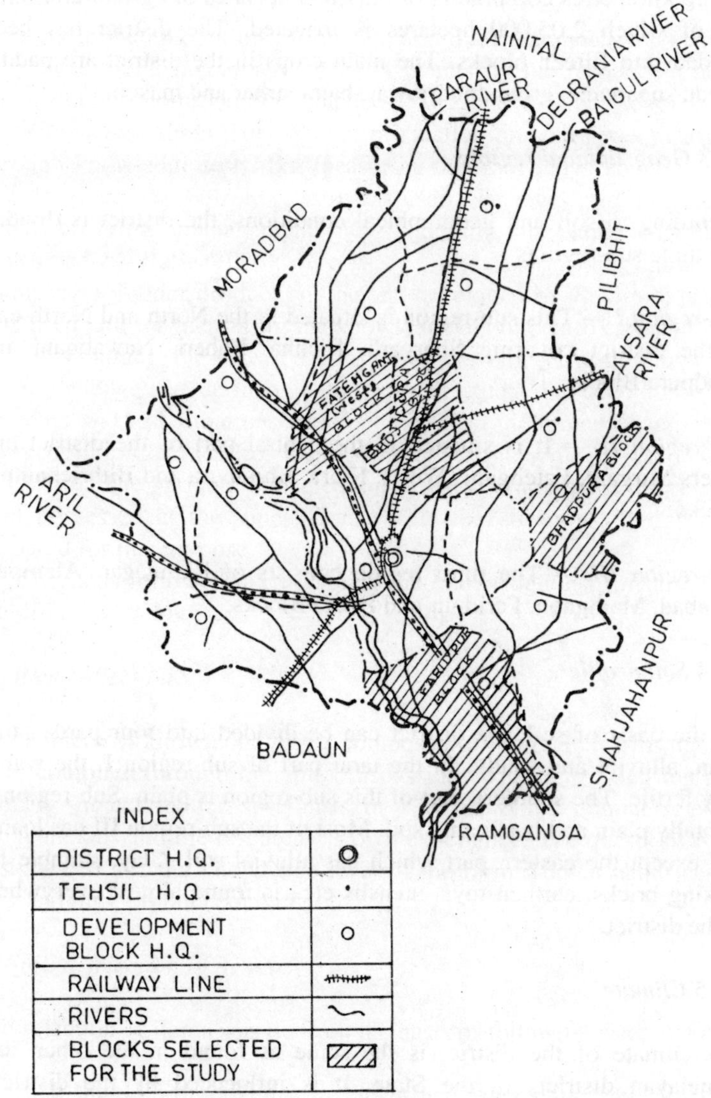

FIG: 3.1 Map of District Bareilly

The total area of Bareilly district according to 1981 census is 4,120 square kilometres comprising of 3,31,000 hectares of agricultural land, out of which 2,05,000 hectares is irrigated. The district has been divided into fifteen blocks. The main crops in the district are paddy, wheat, sugarcane, groundnut, jowar, bajra, arhar and masoor.

3.1.3 *Geographical regions*

According to soil and geographical conditions, the district is divided into three sub-regions.

Sub-region I — This sub-region is situated in the North and North-east of the district covering Shergarh, Richha, Baheri, Nawabganj and Bhadpura Blocks.

Sub-region II — It is situated in the central part of the district and covers Mirganj, Fatehganj (West), Kiara, Bhojipura and Bithrichainpur Blocks.

Sub-region III — The third region consists of Ramnagar, Alampur-Jafrabad, Majhgaon, Faridpur and Bhuta Blocks.

3.1.4 *Soil profile*

On the basis of soil, the district can be divided into four parts: tarai, plain, alluvial and sandy. In the tarai part of sub-region I, the soil is very fertile. The southern part of this sub-region is plain. Sub-region II is totally plain and has loamy soil. Most of the sub-region III has loamy soil except the eastern part which has alluvial soil. Clay, suitable for making bricks, earthen toys, utensils etc., is found almost everywhere in the district.

3.1.5 *Climate*

The climate of the district is the same as found in the other sub-Himalayan districts in the State. It is influenced by the district's proximity to the hills and the Tarai swamps in the north. There are four seasons: the winter season is from December to February, followed by the summer which continues till about the middle of June. The South-West monsoon then ushers in the rainy season which lasts till the end of September. October and November constitute the post-monsoon season.

3.1.6 *Rainfall*

The average rainfall of the district is 1107 mm with the maximum rainfall being recorded in the month of July.

3.1.7 *Temperature*

The mean daily minimum and maximum temperatures of the district are 6.5°C and 44.6°C, respectively.

3.1.8 *Humidity*

The atmosphere is very humid during the monsoon season. The summer season is the driest part of the year when the humidity in the afternoon can be as low as 20 per cent.

3.1.9 *Demographic Information*

Bareilly district has 1901 villages, 1326 Gram Sabhas, 144 Village Panchayats, 5 Tehsils and 15 Blocks. It has 26 Veterinary Hospitals, 26 Livestock Development Centres, 42 A.I. Centres and 83 Cooperative Societies.

The total population of the district, according to 1981 census, is 22.73 lakh, comprising of 12.42 lakh males and 10.30 lakh females, out of which 6.59 lakh were in urban areas and the remaining 16.14 lakh in the rural areas. The density of the population is 552 persons per sq. km. The distribution of urban population is almost equal in all the three sub-regions of the district. Out of the total population of the district 71 per cent live in rural area.

The scheduled castes population is 12.40 per cent of the total population of the district. Occupation-wise, a total of 3,84,932 (16.93%) are engaged in farming, 17,000 (0.74%) in farm labour and 1,40,873 (6.19%) in cottage industries.

Religion-wise, 71.95 per cent of the population in the district is Hindus, 27.15 per cent Muslims and 0.90 per cent belong to other religions.

The male-female ratio is 971, 858 and 870 females per 1000 males in sub-region I, II and III respectively, over-all ratio for the district being 830 females per 1000 males.

The overall literacy percentage of Bareilly district is 22.40. The male literacy is 30.11 per cent, whereas female literacy is 12.33 per cent.

Drinking water facilities are available in 854 villages. The number of villages connected by road is 247, while the number of electrified villages of the district is 1,064.

The district has a total of 70,691 cows above 3 years of age. The female cross-bred cows above 2-$^{1}/_{2}$ years of age are 8,906. The total population of female buffalo above 3 years is 1,30,444.

3.1.10 *Selection of Blocks*

For the selection of the blocks, information about the animal population of all the 15 blocks was collected. The blocks were then divided into two groups — high and low cattle and buffaloes population. Bhadpura and Faridpur blocks from amongst the high population group and Bhojipura and Fatehganj (West) from the low population group were randomly selected, using random tables considering each block as the stratum. A brief description of these blocks is given below:

Bhadpura

Bhadpura block covers an area of 24,198 hectares. The main crops being wheat, gram, paddy and sugarcane. The block has 15,902 households with a population of 90,182. The block has a literacy rate of 17.90 per cent, out of which only 4.74 per cent females are literate. It has a branch post office, a junior high school and a junior basic school for boys, a junior basic school for girls and a cattle pond (Fig. 3.2).

Faridpur

Faridpur, the headquarters of the Tehsil of the same name, has a population of 98,976 in 17,278 households. It has about 16.81 per cent literacy rate out of which 5.03 per cent females are literate. It is spread over an area of 32,177 hectares. The main crops are wheat, gram, bajra, paddy, sugarcane and groundnut. It has an allopathic dispensary, a primary health centre, a family planning centre, a maternity centre, a junior high school, a higher secondary school and a higher secondary school for girls, a posts and telegraphs office, a police station, an inspection house and a cattle pond (Fig. 3.3).

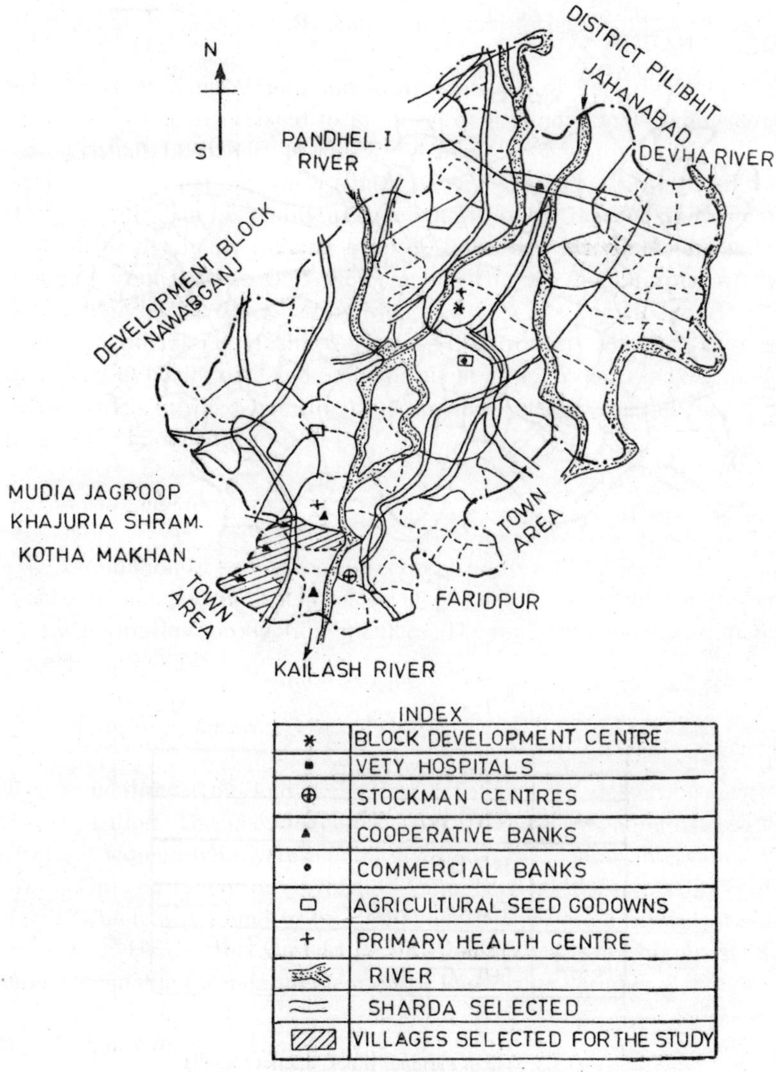

FIG 3.2: Map of Bhadpura Block, district Bareilly

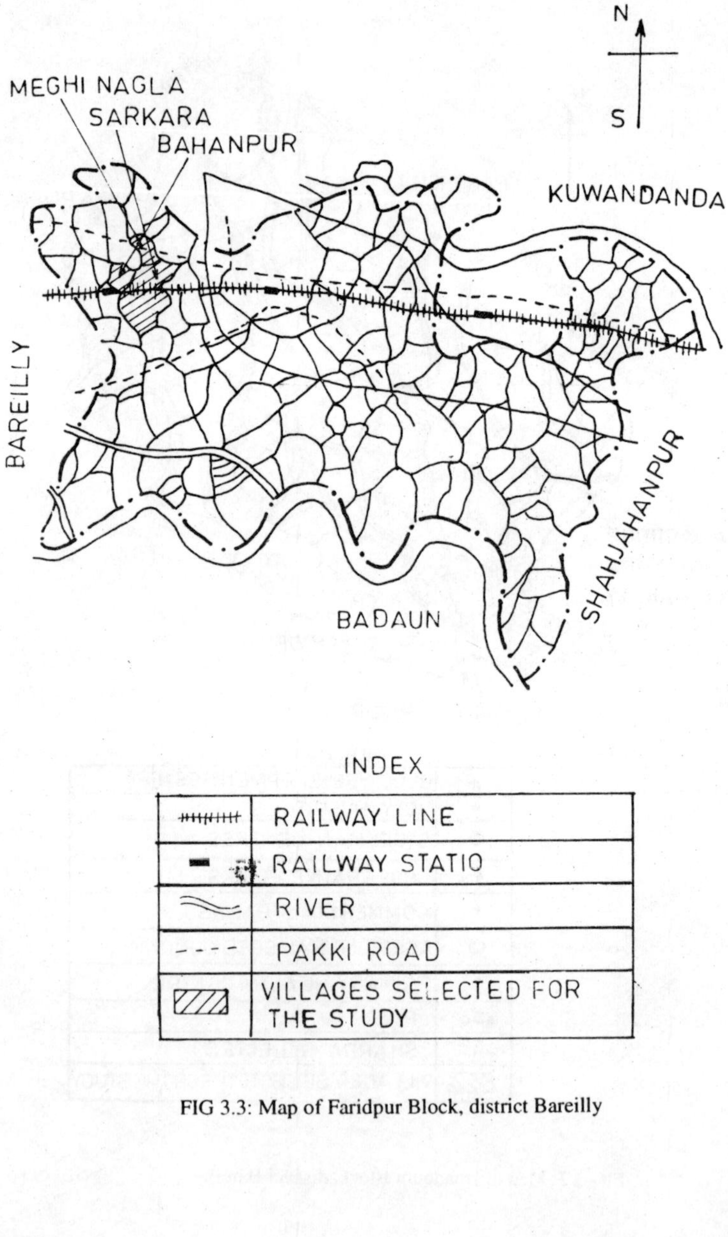

FIG 3.3: Map of Faridpur Block, district Bareilly

Bhojipura

Bhojipura has 14,522 households and an area of 19,661 hectares. The main crops are wheat, gram, paddy, groundnut and sugarcane. The place has a training-cum-production centre where vocational training to carpenter, machinist, fitter-cum-mechanic, electrician and tin-smith is given. A fruit preservation unit has also been started by Kissan Products Ltd. It has a population of 89,093. It has about 14.79 per cent literacy rate, out of which only 4.33 females are literate.

It has an allopathic dispensary, a family planning centre, a maternity centre, a veterinary hospital, a junior high school for boys and a junior basic school each for boys and girls, a cattle pond, a police station and a post office (Fig. 3.4).

Fatehganj (West)

It has a population of 16,027 households and a population of 97,549. It covers an area of 20,056 hectares. It has about 14.27 literacy rate, out of which only 3.95 per cent females are literate (Fig. 3.5).

3.1.11 *Selection of Villages*

Considering each block as stratum, 3 villages were randomly selected from each of the 4 randomly selected blocks. Thus, a total of 12 villages were selected for the study as shown in Table 3.1.

TABLE 3.1: Randomly selected blocks and villages

Blocks		Villages
BHADPURA	1.	Makhan Kotha
	2.	Khajuria Shriram
	3.	Mudia Jagrup
FARIDPUR	1.	Meghi Nagla
	2.	Bahanpur
	3.	Sarkara
BHOJIPURA	1.	Khataula
	2.	Quasampur
	3.	Bibiapur
FATEHGANJ (WEST)	1.	Shahi
	2.	Lamkan
	3.	Vikrampur

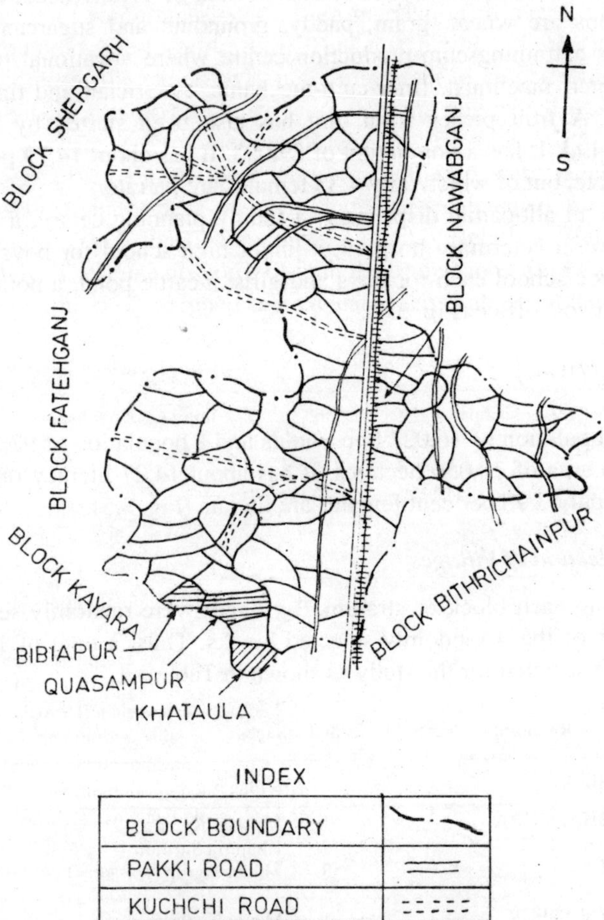

FIG 3.4: Map of Bhojipura Block district Bareilly

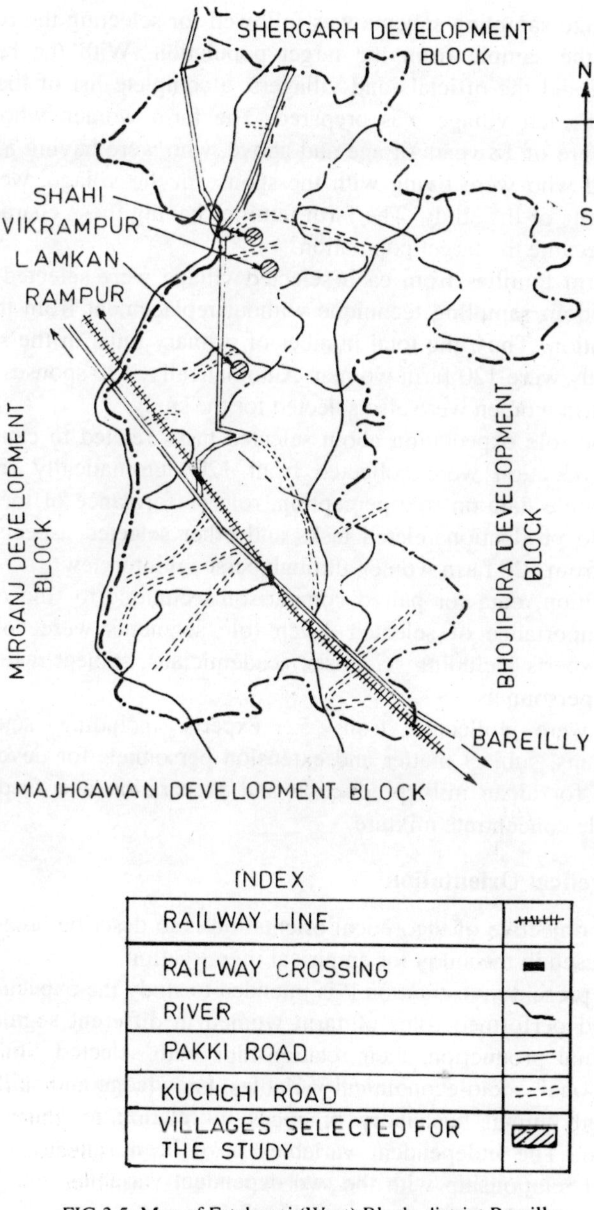

FIG 3.5: Map of Fatehganj (West) Block, district Bareilly

3.1.12 *Selection of Respondents*

A multi-stage sampling scheme was followed for selecting the respondents for the sample from the target population. With the help of voters' list and the officials and villagers, a complete list of the farm families of each village was prepared. The farm women who were married, were of 18 years of age and above, who were having a milch animal and who were living with the spouse in the village, were the primary units of the study. The farm families having these characteristics thus became the target population.

Ten farm families from each selected village were selected using simple random sampling technique without replacement from the target population. Thus, the total number of primary units in the sample for the study were 120 farm women. Automatically, 120 spouses of the selected farm women were also selected for the study.

Data on role expectation about selected tasks related to cattle and buffalo production were collected from 120 automatically selected spouses, while data on role perception, role performance of the cattle and buffalo production related tasks and other selected aspects were collected from 120 farm women through personal interview.

In addition, data for paired comparison technique to find out the relative importance of selected seven role segments were collected from 50 experts including scientists, academicians, subject matter and extension personnels.

Data were collected from 52 experts including scientists, academicians, subject matter and extension personnels for developing skill tests for clean milk production and preparation and feeding of home made concentrate mixture.

3.2 **Theoretical Orientation**

The main objective of theoretical orientation is to describe broadly the concepts used in the study for empirical investigation.

In the present investigation it is intended to study the expected, perceived and performed roles of farm women in different segments of dairy animal production, their relationships with selected situational, socio-personal, socio-economical variables, knowledge and skill levels of selected animal husbandry practices in relation to dairy animal production. The independent variables have been selected on their theoretical relationship with the two dependent variables viz. productivity of milk and role performance (Fig. 3.6).

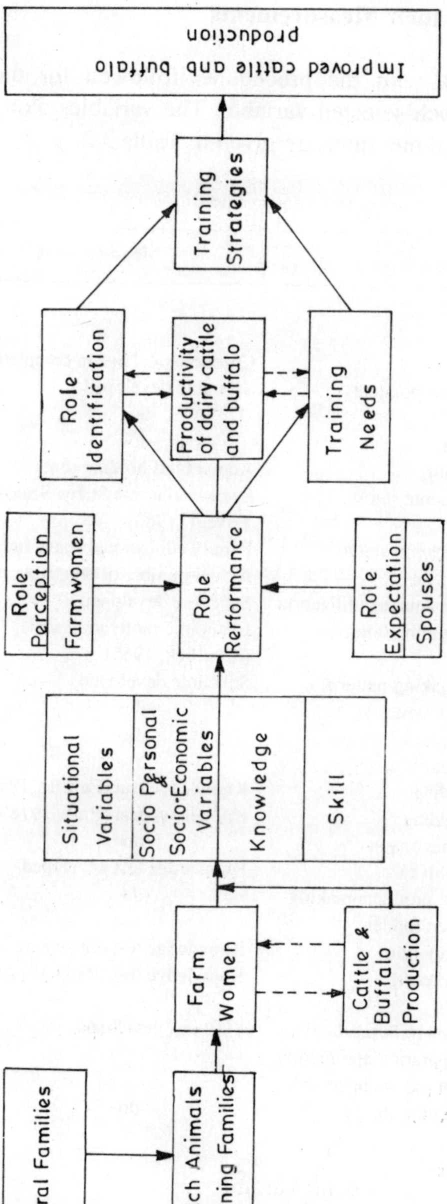

FIG 3.6: Conceptual model for role dynamics of farm women in relation to productivity of dairy cattle and buffalo

3.3 Variables and their Measurements

This section deals with the procedures followed for the empirical measurement of each selected variable. The variables along with their respective empirical measures are given in Table 3.2.

TABLE 3.2: Variables studied and measures used

	Variables	Measurements
	Independent	
X_1	Age	Chronological age in completed years
X_2	Position of respondent	Schedule developed
X_3	Caste	-do-
X_4	Family type	-do-
X_5	Land-holding	Actual land-holding (ha)
X_6	Socio-economic status	Socio-economic Status Scale (Rural) Trivedi (1963)
X_7	Family education status	Family educational status index (Ray, 1967)
X_8	Dairy herd size	Actual number of breedable milch animals
X_9	Information sources utilization	Schedule developed
X_{10}	Economic motivation	Economic motivation scale (Moullick, 1965)
X_{11}	Decision-making pattern	Schedule developed
X_{12}	Role expectation	-do-
X_{13}	Role perception	-do-
X_{14}	Knowledge:	
	i. Feeding	Knowledge test (Fulzele, 1986)
	ii. Breeding	Knowledge test (Rath, 1978)
	iii. Management	-do-
	iv. Health care	Knowledge test developed
	v. Milk products making	-do-
	vi. Green fodder production	Knowledge test (Goswami, 1987)
	vii. Marketing	Knowledge test (Rath, 1978)
X_{15}	Skill:	
	i. Clean milk production	Skill test developed
	ii. Preparation and feeding of home made concentrate mixture	-do-
	Dependent	
Y_1	Productivity of milk	Livestock production index (Milk Yield), (Yang, 1980)
Y_2	Role performance	Schedule developed

3.4 **Empirical Measurement of Variables**

The methods and procedures for the measurement of each variable are given below:

3.4.1 *Independent Variables*

i. *Age*

Age was operationalized as the number of years completed by the respondent at the time of enquiry. In the present study the respondents were categorized as:

Old	:	Mean + S.D.
Middle aged	:	Mean ± S.D.
Young	:	Mean – S.D.

ii. *Position of Respondents*

It was operationalized as the status of respondent in the family as housewife, daughter-in-law and mother-in-law and the scores given were 1, 2 and 3 respectively.

iii. *Land-holding*

Land-holding refers to the total land in hectares, possessed by the family of the respondent. The information regarding land owned was obtained as a part of socio-economic status scale (Trivedi, 1963).

iv. *Family type and size*

Family type refers to nuclear or joint family. Family size refers to the total number of family members living under the same roof. Big size of family means, more expenses towards family budget and more need of earning to cope up with the expenditure. This was also operationalized according to socio-economic status scale of Trivedi (1963).

v. *Socio-economic status*

This refers to the position and respect of the respondent enjoyed in her own limited society as determined by wealth, occupation and social class. Its measurement was done according to the socio-economic status categories given by Trivedi (1963), as upper class (43 scores and above), upper-middle class (32 to 42 scores), middle class (21 to 31 scores), lower-middle class (10 to 20 scores) and lower class (below 10 scores).

Scoring technique of socio-economic status: The quantification of all the aspects of socio-economic status except age and position of respondent, is based on socio-economic status scale developed by Trivedi (1963). The measurement of individual item is given in the scale. The socio-economic status score of each respondent was obtained by adding all the scores of individual items of socio-economic status. The different categories of socio-economic status were made according to total scores obtained by the respondents.

vi. *Dairy herd size*

It refers to the total number of breedable cows and buffaloes owned by the family of the respondent.

vii. *Family educational status*

It refers to academic qualifications of all the members acquired through formal schooling. The years of schooling of different family members were quantified by using socio-economic status score (Trivedi, 1963) and then overall family education status of the family was computed by using the formula used by Ray (1967):

$$\text{Family Education Status Index} = \frac{\text{Total education score of family}}{\text{Effective family size}}$$

where, effective family size is the number of family members above four years of age.

The respondents were categorised according to family educational status as:

High	:	Mean + S.D.
Medium	:	Mean ± S.D.
Low	:	Mean – S.D.

viii. *Economic motivation*

It is occupational success in terms of profit maximization and relative value an individual places on economic ends. In the present study economic motivation referred to one's inner desire to maximize production as well as profit from livestock enterprise. It was measured by the self-rating scale developed by Moullick (1965). The obtained score ranged from 1 to 9.

The computation of economic motivation was done as under:

$$\frac{\text{Most liked}}{\text{Least liked}}$$

ix. *Information sources utilization*

This refers to the sources/channels of information which add to the knowledge of respondents (livestock owners). For quantifying the sources/channels of information used by respondents, sixteen institutionalised and non-institutionalised information sources relevant to animal husbandry practices were listed. To measure the degree of information sources utilization, each respondent was asked to indicate, on a 4-point continuum, as to how often she (farm women) got the information about animal husbandry practices from each of the listed sources. The scoring procedure for the respondents was: Most often: 3, Often: 2; Sometimes: 1 and Never: 0.

x. *Decision-making pattern*

It referred to the extent of participation in decision-making process by farm women about the tasks associated with dairy animal production in which the family is engaged in. Through the structured schedule, the decision-making pattern of the family was measured and the scores assigned were as follows:

> Spouse only:1, All family members/friends/relatives:2, House-wife + Spouse:3 and House-wife only:4.

xi. *Role expectation*

Role expectation constitutes the ways of behaving which are expected of any actor occupying a certain position. Role expectations are considered by most role theorists to be the essential ingredient in any formula of predicting role behaviour. Role expectation is, therefore, operationalized as what an individual is expected to be or how an individual is expected to behave as an incumbent of a particular position in a group or sub-system or system. The expectation of roles of farm women by their respective spouses was measured on a 5-point continuum; Must be done by house-wife:5, Should be done by house-wife:4, May be done by either house-wife or the spouse:3, Should not be done by house-wife:2 and Must not be done by house-wife:1.

The role expectation of each sub-role has been measured on 5-point continuum of which the middle point i.e. 3 is the null point. Therefore, the respondents (spouse) were categorised according to role expectation as having:

High	:	above 3 mean score
Medium	:	3 mean score
Low	:	less than 3 mean score

xii. *Role perception*

In the present study the role perception means the functions or activities that a person occupying a given position THINKS or BELIEVES should be performed as duties/tasks.

On the basis of expectations of related incumbents, the individual makes his/her own perceptions in term of what role should be performed to justify his/her given status. The perceived role of farm women is, therefore, defined in terms of what they themselves think of their own role in relation to the studied animal husbandry tasks. The perception of role, as perceived by farm women by themselves, was measured on a 5-point continuum; Must be done by house-wife only:5, Should be done by house-wife:4, May be done by either house-wife or the spouse:3, Should not be done by house-wife:2 and Must not be done by house-wife:1.

The respondents (farm women) were categorized according to role perception as having

High	:	above 3 mean score
Medium	:	3 mean score
Low	:	less than 3 mean score

xiii. *Knowledge*

It refers to the amount of information and understanding which can be recalled through memory of ideas or events by the respondent about cattle and buffalo production practices.

Knowledge about dairy animal production practices was measured through knowledge tests. For measuring the knowledge about feeding test developed by Fulzele (1986) was used. For measuring knowledge about breeding, management and marketing, the knowledge tests developed by Rath (1978) were used. For measuring the knowledge about green fodder production, the test developed by Goswami (1987) was used and for measuring the knowledge about health care and milk products making, the knowledge tests were developed for the study.

Knowledge, as defined in the present study, included those behaviours and test situations which emphasized the remembering either by recognition or recall of ideas, material or phenomenon (Bloom, *et al.*, 1956). The variable indicated the extent of knowledge the respondent possessed at the time of answering the questions scientifically prepared for this purpose.

The steps followed in developing the knowledge test for this study are discussed below:

(a) *Item collection*

The content of knowledge test is composed of items. Items for the test were compiled through literature, discussions with field extension personnel, subject matter specialists in veterinary and animal science, academicians and the researcher's own experience. The questions were designed to test the knowledge level of the farm women about health care and milk products making.

(b) *Initial selection of items*

The selection of items was done on the basis of following criteria:

* It should promote thinking rather than role memorization, and
* It should differentiate the well informed farm women from the poorly informed ones and should have a certain difficulty value.

The procedure followed in selection of the test items was on the lines used by Choudhari (1978), Sagar (1983) and Goswami (1987). Based on these two criterion, 180 items were initially constructed out of which 30 were about health care and 150 were about different milk products making. All the 180 items collected for construction of the knowledge test were in objective form and were in dichotomous or multiple choice format.

A schedule was prepared with these 180 items for administering to the farm women for item analysis and screening out further items.

(c) *Item analysis*

The item analysis of a test usually yields two kinds of information; item difficulty and item discrimination. The index of item difficulty reveals how difficult an item is, whereas the index of discrimination indicates the extent to which an item discriminates the well informed individual from the poorly informed ones.

The items were checked and modified on the basis of pretesting and administered to 60 farm women owning cattle and buffaloes for item analysis. The farm women owning cattle and buffaloes were randomly selected for administering the item and were different from the sample for the present study. Nevertheless, these 60 farm women were representative of the community in which the final study was conducted.

Each one of the 60 respondents, to whom the test was administered, was given score 1 or 0 for each item according to whether the answer was right or wrong. The total number of correct answers given by farm women out of 180 items was the knowledge score secured by her. After calculating the scores obtained by 60 farm women, the scores were arranged in a descending order.

These 60 farm women were then divided into six equal groups, each having 10 farm women. Farm women in each group were arranged in descending order according to the score obtained by each one of them. These groups were named as G_1, G_2, G_3, G_4, G_5 and G_6, respectively. For item analysis, the middle two groups, that is G_3 and G_4 were eliminated. Only four extreme groups with high and low scores were considered for computation of item difficulty and their discrimination indices.

The range of the scores obtained by the four groups of respondents were as follows:

G_1 : 172 to 154

G_2 : 151 to 128
G_5 : 106 to 88
G_6 : 84 to 65

The maximum score was obviously 180 which could be scored when all the 180 items were answered correctly. The data pertaining to the correct answers from the items in respect of these four groups were tabulated.

(d) *Calculation of difficulty index*

The difficulty index of an item was defined as the proportions of farm women giving correct answers to that particular item. This was calculated by the formula:

$$P_i = \frac{n_i}{N_i} \times 100.$$

where,

P_i = Difficulty index in percentage of *i*th item,
n_i = Number of farm women giving correct answer, to *i*th item,
N_i = Total number of livestock owners to whom *i*th item was administered.

(e) *Calculation of discrimination index*

The discrimination index was obtained by calculating the phi-coefficient as formulated by Perry and Michael (1951). However, Mehta (1958) while using E 1/3 method to find out item discrimination emphasised that this method was analogous to, and hence, a convenient substitute for the phi-coefficient. The method suggested by Mehta (1958) was adopted for the present study. The formula by which the item discrimination index was calculated is given below:

$$E\ 1/3 = \frac{(S_1+S_2) - (S_5+S_6)}{N/3}.$$

where,

S_1, S_2, S_5 and S_6 = The frequencies of correct answers in G_1, G_2, G_5 and G_6 groups, respectively.

N = Total number of livestock owners in the sample of item analysis.

The discrimination indices of all the items calculated by the procedure mentioned above.

An example of computation of difficulty and discrimination indices of knowledge items is given below:

TABLE 3.3: Difficulty and discrimination indices of knowledge items

Sl.No.*	Frequencies of correct answers				Total Frequencies	P** (N=60)	E 1/3
	S_1	S_2	S_5	S_6			
A.5	10	8	6	4	45	75.00	0.40
A.17	10	10	8	9	54	90.00	0.15
B.i.11	9	7	7	2	36	60.00	0.35
B.i.28	6	9	7	5	44	73.33	0.15
B.iii.10	10	8	6	4	41	68.33	0.40
B.iv.17	10	10	7	6	48	80.00	0.35
B.iv.35	10	10	10	8	56	93.33	0.10

* Serial number of items in the initial test.

** Percentage of respondents giving correct answers.

Substituting the values for item number B.i.28 of the above illustration the value arrived at :

$$P_i \text{ (Difficulty index)} = \frac{n_i}{N_i} \times 100$$

$$= \frac{44}{60} \times 100$$

$$= 73.33$$

$$E1/3 \text{ (Discrimination index)} = \frac{(S_1+S_2) - (S_5+S_6)}{N/3}$$

$$= \frac{(6+9) - (7+5)}{20}$$

$$= \frac{3}{20}$$

$$= 0.15$$

(f) *Selection of items for test*

Two criteria viz. item difficulty and item discrimination index were considered for selection of items in the final format of the knowledge test.

The underlying assumption in the statistics of item difficulty was that it was linearly related to the level of the individual's knowledge about animal husbandry practices. When a respondent gave the correct response to an item, it was assumed (Coombs, 1950), that the item was less difficult than his ability to cope with it. In the present study, items with difficulty index ranging from 30 to 80 and discrimination index ranging from 0.30 to 0.55 were included in the final format of the knowledge test as described by Mehta (1958). The knowledge test had two components i.e. Health Care and Milk Products Making. Thirteen items about health care and 64 items about milk products making practices, which fulfilled both the criteria, were selected for the final format of the knowledge test.

(g) *Scoring method*

The summation of scores for correct replies over all the items of a particular respondent indicated his level of knowledge about the studied cattle and buffalo production practices. The range of scores was, therefore, from 0 to 77.

(h) *Reliability of knowledge test*

The reliability of the knowledge test developed was tested by Test – retest method. The knowledge test with 77 items was administered to 30 farm women who were owning livestock, twice at an interval of 15 days. The coefficient of correlation value between the two scores was 0.879 which was found to be significant at one per cent level of significance. Hence, the knowledge test constructed, was highly stable and dependable for measurement of the knowledge variable.

(i) *Content validity of knowledge test*

In the final selection of items, care was taken to include items covering the entire universe of relevant behavioural aspects of the respondents with respect to knowledge about health care and milk products making practices. Items were collected through various sources including specialists of the concerned subject matter and it was assumed that the scores obtained by administering the tests measured knowledge of the respondents as it was intended.

(j) *Quantification of knowledge*

On the basis of these scores obtained by a respondent, the knowledge level was calculated in terms of percentage. The scores obtained by a respondent for a particular segment were divided by the maximum obtainable scores of that role, multiplied by 100. The knowledge of the respondents was categorized into various levels as:

High	:	60 per cent and above
Medium	:	30-59 per cent
Low	:	0-29 per cent

(xiv) *Skill*

Skill is the complex organisation of physical or verbal behaviour developed through learning and directed towards a particular goal or centred on a specific activity. Mace (1950), Gatharie (1952) and King (1960) have defined skill in terms of ability to perform a given task accurately. Skill in the present study has been operationalized in relation to the studied animal husbandry practices.

The skills of farmers and farm women have been measured by many researchers on the basis of expressed opinions. These expressed opinions, based entirely on self perception, are sometimes likely to become subjective. A farm woman may perceive herself performing a given task correctly, but the procedure actually followed in doing that task may not be according to the scientific recommendations. Such farm women may get high scores for skill competence although they may not be so competent when objectively assessed on the performance of that skill. Keeping this in view, the two major jobs/practices in relation to cattle and buffalo production were identified which are performed mostly by farm women. This was done on the basis of available literature, discussions and by general observations in the area of study.

The jobs/practices selected for testing the skill of farm women were clean milk production and preparation and feeding of home made concentrate mixture. For testing the skill, the procedure followed by Weirsma and Jurs (1976), Singh (1981) and Fulzele (1986) for measuring psychomotor variables, was used as a guideline, with some modifications, for the development of skill tests for the two selected jobs/practices.

The jobs/practices were considered as skilled if they were both scientific and practical and had at least ten recognizable and observable sequential steps while being performed on the basis of scientific recommendations.

(a) *Selection of the Steps*

All the steps were listed in a systematic and sequential order with the help of available literature and detailed consultations with the scientists, academicians and specialists in the concerned subject matter areas. The mutually exclusive steps in the developed skill tests will become "observation points" at the time of actually measuring the competence of the performer of the skill, in terms of accuracy, as per the scientific recommendations.

(b) *Scoring Procedure*

After listing the steps in a manner that each step was clearly identifiable, observable and mutually exclusive of the other listed steps, the lists of steps in clean milk production and preparation and feeding of home made concentrate mixture were referred to 52 scientists, academicians, subject matter specialists and field extension officers from Indian Veterinary Research Institute, Izatnagar, G.B. Pant University of Agriculture and Technology, Pantnagar, College of Veterinary Science and Animal Husbandry, Mathura, Haryana Agriculture University, Hisar and National Dairy Research Institute, Karnal. They were requested to separately assign scores to each listed step. Each listed step was to be assigned score for its "Difficulty Index" and "Relative Importance", separately so that the total of scores assigned to all the listed steps was 50 for each of the two dimensions.

"Difficulty Index", for the purpose of this study, was defined as the extent of difficulty the performer of the skill is likely to experience in performing the step in relation to every other listed steps required to perform the skill. Thus a step which, in the judgement and opinion of the expert, was likely to be more difficult to perform by the farm women, was given higher score than the one which was likely to be less difficult to perform, as compared to other listed steps.

"Relative Importance", for the purpose of this study, was defined as the degree of importance of the step in relation to every other listed step required to perform the skill. Thus a step which, in the judgement and opinion of the expert, was more important to be performed by the farm women, was given higher score than the one which was less important, as compared to other listed steps.

Scores given by all the experts to each step were tabulated and mean scores, in respect of each step, were calculated for both the dimensions i.e. "Difficulty Index" and "Relative Importance", separately. These mean scores of both the dimensions were then pooled for getting the final assigned score to each step (Tables 3.4 and 3.5).

TABLE 3.4: Mean and pooled scores of recommended steps for clean milk production

	Recommended Sequential Steps	Mean scores on difficulty index (N=52)	Mean scores on relative importance (N=52)	Pooled mean scores (N=52)
1	2	3	4	5
1.	Selecting milking utensil having a narrow mouth	1.39	1.37	2.76
2.	Selecting proper size of milking utensil	1.24	1.18	2.42
3.	Selecting flat bottomed milking pot to hold it properly while milking	1.21	1.21	2.42
4.	Cleaning tho milking pot	1.18	2.52	3.70
5.	Selecting clean, quiet place for milking	1.56	1.76	3.32
6.	Milking the animal at proper interval	2.12	2.26	4.38
7.	Keeping weaned animal away from the milked animal	1.81	1.55	3.36
8.	Leaving the calf to the animal for suckling before starting to milk	1.10	1.30	2.40
9.	Taking away calf and tying it separately as it starts suckling	2.05	1.27	3.32

Contd.

Table 3.4 (*Contd.*)

1	2	3	4	5
10.	Providing conc. mixture/feed to the animal while milking	2.17	2.96	5.13
11.	Tying the hind legs of the milked animal above stifle joint	1.40	1.12	2.52
12.	Tying the tail of the milked animal	1.34	1.01	2.35
13.	Using proper rope with correct knot for tying the hind legs	1.29	1.32	2:61
14.	Using clean water for washing etc.	1.25	2.18	3.43
15.	Washing the hands with soap and water	1.46	1.89	3.35
16.	Washing the udder and teats with soap and water	1.91	2.19	4.10
17.	Using clean cloth for drying udder and teats	1.49	1.62	3.11
18.	Wiping hands with clean cloth	1.21	1.43	2.64
19.	Milking from left side of the animal	1.74	1.23	2.97
20.	Sitting for milking at proper distance	1.32	1.34	2.66
21.	Holding the utensil properly for milking	1.75	1.19	2.94
22.	Observing few milk drops for any abnormality in the milk	1.82	2.20	4.02
23.	Milking two diagonal teats at a time	1.74	1.84	3.58
24.	Using both the hands in palm milking	2.83	1.80	4.63
25.	Milking the animal in one sitting	1.69	1.74	3.43
26.	Releasing the calf for feeding of the milk left in the udder	1.28	0.89	2.17
27.	Pulling the end of rope to untie the legs	1.12	0.86	1.98
28.	Taking precautions so that animal does not get hurt	1.42	1.25	2.67
29.	Covering the utensil containing milk properly	1.41	1.40	2.81
30.	Transporting the milk to a safe, clean place	1.35	1.25	2.60

Contd.

Table 3.4 (*Contd.*)

1	2	3	4	5
31.	Filtering milk through clean piece of cloth	1.91	1.47	3.38
32.	Storing milk in cool temperature or Boiling the milk for storage	1.44	1.40	2.84
	Total	50	50	100

TABLE 3.5: Mean and pooled scores of recommended steps for preparation and feeding of home made concentrate mixture

	Recommended Sequential steps	Mean of scores on difficulty index (N=52)	Mean of scores on relative importance (N=52)	Pooled mean scores (N=52)
1.	Assessing the quality of available ingredients	7.08	7.21	14.29
2.	Selecting the ingredients	5.25	6.37	11.62
3.	Cleaning the ingredients	3.93	3.21	7.14
4.	Breaking the oil cake into small pieces before grinding	3.14	2.79	5.93
5.	Grinding the ingredients	4.01	3.72	7.73
6.	Measuring different ingredients in required quantities	4.40	4.49	8.89
7.	Cleaning the feeding trough/ floor for mixing the ingredients	2.88	2.86	5.74
8.	Selecting the proper utensil for soaking the conc. and straw.	2.74	2.42	5.16
9.	Cleaning the utensil used for soaking	2.63	2.54	5.17
10.	Mixing different ingredients in proper proportion	3.82	3.76	6.37
11.	Mixing the required quantities of conc. and straw in the utensil	3.14	3.23	6.37
12.	Adding the required quantity of water in the utensil to soak the conc. mixture and straw completely	2.43	2.65	5.08
13.	Soaking of mixed feed for 6-8 hours	2.16	2.48	4.64
14.	Mixing the soaked feed again before offering to the animals	2.39	2.27	4.66
	Total	50	50	100

(c) *Reliability and validity of the test*

The reliability of the skill tests for clean milk production and preparation and feeding of home made concentrate mixture was computed with the help of split-half method and was found to be 0.92 and 0.90 respectively. The following formula was used (Singh, 1981):

$$rtt = 1 - \frac{\sigma^2 d}{\sigma^2 t}$$

where,

rtt = Reliability of the total test
d = Difference between two half scores
$\sigma^2 d$ = Variance of these differences
$\sigma^2 t$ = Variation of the total scores

The intrinsic validity of the test was calculated by taking square root of the reliability coefficient which was 0.96 for clean milk production and 0.95 for preparation and feeding of home made concentrate mixture, thus indicating high reliability and validity of the tests. The content and concurrent validity for the skill tests were ensured through jury validation.

The categorization of respondents into different skill levels was done as:

High	:	60 per cent and above
Medium	:	30 to 59 per cent
Low	:	0 to 29 per cent

3.4.2 *Dependent Variables*

i. *Productivity of Milk*

Productivity has been conceptualized as a manifestation of obtaining total fluid milk from per unit of dairy animal through the adoption of scientific dairy practices, available manpower and financial resources.

Milk production of dairy animals (cattle and buffaloes), which was taken as the dependent variable in this study, was measured by the Livestock Production Index (Milk Yield) as recommended by Yang (1980). According to him milk production index represents the average milk yield of all the dairy animals (cattle and buffaloes) on a dairy farm compared with the average milk yield of the region. Before calculating the milk production index for the milch animals of a particular dairy farm (respondent), the average yield of the milch animal (cow/buffalo) in the region has to be determined. Then, by dividing the average yield per cow/buffalo on the particular dairy farm (respondent) by the average milk yield of a cow/buffalo in the region, a percentage figure is obtained which, when multiplied by 100 gives the milk index number for the individual animal.

For the purpose of the present study four milch animals of the area i.e. cross-bred cow, desi cow, improved buffalo and desi buffalo were taken into consideration. The average milk yield of each of these dairy animals, for the year preceding the year of study, in respect of each respondent was recorded. Average milk yield of each dairy animal over the 120 respondents was taken as the average milk yield of that milch animal for the region.

The categorization of productivity level was done as:

High	:	Mean + S.D.
Medium	:	Mean ± S.D.
Low	:	Mean – S.D.

ii. *Role Performance*

Role performance was defined as the actual behaviour of an incumbent in a position. According to Davis (1949) role is how an individual actually performs in a given position, as distinct from how that individual is supposed to perform.

In this study, role performance was measured by the actual behaviour of the farm women which has relevance to the pursuance of her work in cattle and buffalo production as perceived by her. The role performance of farm women, was measured on the following 5-point continuum; Performed very well:5, performed well:4, performed averagely:3, performed poorly:2, and performed very poorly:1.

The categorization of role performance (farm women) was done considering 3 as null point, into different categories as:

High	:	Mean score more than 3
Medium	:	Mean score 3
Low	:	Mean score less than 3

3.4.3 *Relative Importance of Role Segments in Dairy Animal Production*

In the present investigation, the paired comparison technique was used to find out the relative importance of the seven major role segments in dairy animal production viz. feeding, breeding, management, health care, milk products making, green fodder production and marketing. This was considered necessary in order to identify the relative importance of each segment. This method of comparing each role segment with every other role segment facilitated in assigning weightage, as per its importance, in the overall expected, perceived and performed role of the farm women in dairy animal production.

The procedure followed in this study was as recommended by Edward (1958). According to the requirements of the paired comparison technique, the seven major role segments of the farm women being studied, were presented to the respondents in pairs of all possible combinations as per the following formula:

$$\frac{n(n-1)}{2} = \frac{7(7-1)}{2} = 21$$

where,

n = Total number of segments

These 21 combinations were sent to 50 scientists, academicians subject matter and extension specialists from Indian Veterinary Research Institute, Izatnagar, National Dairy Research Institute, Karnal, G.B. Pant University of Agriculture and Technology, Pantnagar and College of Veterinary Science and Animal Husbandry, Mathura. The experts were acquainted with the procedure of choosing one role over the other from each pair as per the instructions. Responses were received from all the 50 respondents.

i. *Frequency (F) Matrix*

Observed frequencies of choices were arranged in the frequency (F) matrix. This matrix summarised the number of times one stimulus (i) was judged greater or more favourable than other stimulus (j) of the pair (Table 3.6).

TABLE 3.6: F matrix for seven role segments*

Role segment	1	2	3	4	5	6	7
1. Feeding	—	12	15	20	5	13	5
2. Breeding	38	—	35	33	14	24	11
3. Management	35	15	—	21	6	14	4
4. Health care	30	17	29	—	10	18	12
5. Milk products making	45	36	44	40	—	35	22
6. Green fodder production	37	26	36	32	15	—	12
7. Marketing	45	39	46	38	28	38	—
Total	230	145	205	184	78	142	66

* Column segment judged more favourably than row segment.

The segments were arranged in ascending order on the basis of total column frequencies achieved through F-matrix. Table 3.7 gives the F matrix rearranged in ascending order.

TABLE 3.7: *F* matrix in ascending order for seven role segments

Role segment	7	5	6	2	4	3	1
7	—	28	38	39	38	46	45
5	22	—	35	36	40	44	45
6	12	15	—	26	32	36	37
2	11	14	24	—	33	35	38
4	12	10	18	17	—	29	30
3	4	6	14	15	21	—	35
1	5	5	13	12	20	15	—
Total	66	78	142	145	184	205	230

ii. *P Matrix*

From the F matrix (Table 3.7) P matrix was constructed which gave the observed proportion of times stimulus i was preferred to stimulus (j). This was done by dividing each of the cell entries by N i.e. 50 (Table 3.8).

TABLE 3.8: *P* matrix corresponding to the *F* matrix

Segment	7	5	6	2	4	3	1
7	—	0.560	0.760	0.780	0.760	0.920	0.900
5	0.440	—	0.700	0.720	0.800	0.880	0.900
6	0.240	0.300	—	0.520	0.640	0.720	0.740
2	0.220	0.280	0.480	—	0.660	0.700	0.760
4	0.140	0.200	0.360	0.340	—	0.580	0.600
3	0.080	0.120	0.280	0.300	0.420	—	0.700
1	0.100	0.100	0.260	0.240	0.400	0.300	—

iii. *Z Matrix*

From the P matrix, Z matrix was constructed based on the conversion Table I in Edwards (1958) expressing P values into Z values which were units normal deviates corresponding to the proportions. Z values of each stimulus were summed in each column and means were calculated. To make the scales positive, the scale value of the stimulus with largest negative deviation was added to all the scale values. This made the scale value of that stimulus zero and the values of all the others positive. Thus the final scale values were determined by shifting the arbitrary origin to the lowest scale value (Table 3.9).

TABLE 3.9: Z matrix corresponding to P matrix

Segment	7	5	6	2	4	3	1
7	—	0.151	0.706	0.772	0.706	1.405	1.282
5	-.151	—	0.524	0.583	0.842	1.175	1.282
6	-.706	-.524	—	0.050	0.358	0.583	0.643
2	-.772	-.583	-.050	—	0.412	0.524	0.706
4	-1.080	-.842	-.358	-0.412	—	0.202	0.253
3	-1.405	-1.175	-.583	-.524	-.202	—	0.524
1	-1.282	-1.282	-.643	-.706	-.253	-0.524	—
Σ	-5.396	-4.255	-.404	-.237	1.863	3.365	4.690
X	-.771	-.608	-.058	-.034	.266	.481	.670
Scale values	.000	0.163	0.713	.737	1.037	1.252	1.441

For the present study, the actual scale values obtained by the method of paired comparison technique from 50 judges were considered. Based on these scale values the relative importance of seven role segments was obtained and is given in Table 3.10.

TABLE 3.10: Scale values and the relative importance of seven role segments

Role segment	Scale value	Rank order	Weights
Feeding	1.441	1	7
Management	1.252	2	6
Health care	1.037	3	5
Breeding	0.737	4	4
Green fodder production	0.713	5	3
Milk products making	0.163	6	2
Marketing	0.000	7	1

iv. *Internal Consistency Check*

Having obtained the scale values of the seven segments on a least to most favourable psychological continuum, an internal consistency check was applied. This check involved determining how well the observed proportions agree with those to be expected in terms of the desired scale values. The first step was to obtain a Z′ matrix of theoretical normal deviates corresponding to the scale separations of the statements. Table 3.11 gives the corresponding Z′ matrix.

TABLE 3.11: Z′ matrix corresponding to the scale distances between the seven role segments

Segment	Scale values	7	5	6	2	4	3	1
		0	0.163	0.713	0.737	1.037	1.252	1.441
7	0	—						
5	0.163	-0.163						
6	0.713	-0.713	-.0.550	—				
2	0.737	-0.737	-0.574	-0.024	—			
4	1.037	-1.037	-0.874	-0.324	-0.300	—		
3	1.252	-1.252	-1.089	-0.539	-0.515	-0.215	—	
1	1.441	-1.441	-1.278	-0.728	-0.704	-0.404	-0.189	—

Now with theoretical normal deviates (Z′ values), the corresponding theoretical proportions (P′ values) were obtained (Table I, Edwards 1958). Table 3.12 gives P′ matrix.

TABLE 3.12 : P′ matrix corresponding to the Z′ matrix

Segment	7	5	6	2	4	3	1
7	—						
5	0.435	—					
6	0.238	0.291	—				
2	0.231	0.283	0.490	—			
4	0.150	0.191	0.373	0.382	—		
3	0.105	0.138	0.295	0.303	0.415	—	
1	0.075	0.101	0.233	0.241	0.343	0.425	—

The entries in the P′ matrix were substracted from the corresponding entries in the P matrix, that is, the entries of Table 3.12 were substracted from the independent entries of Table 3.8 to obtain the discrepancies between the observed proportions (P) and theoretical proportions (P′). Table 3.13 gives the P – P′ matrix.

TABLE 3.13 : Discrepancies between the P′ values and the P values.

Segment	7	5	6	2	4	3	1
7	—						
5	0.005	—					
6	0.002	0.009	—				
2	–0.011	–0.003	–0.010	—			
4	–0.010	0.009	–0.013	–0.042	—		
3	–0.025	–0.018	–0.015	–0.003	0.005	—	
1	0.025	–0.001	0.027	–0.001	0.057	–0.125	—
Σ	0.078	0.040	0.065	0.046	0.062	0.125	

Taking the sum of the absolute values and dividing by the number of discrepancies, the absolute average discrepancy (AD) was obtained by the following formula:

$$AD = \frac{\Sigma(P - P')}{\frac{n(n-1)}{2}} = \frac{0.416}{21} = .0198 = 0.02$$

The calculated AD values for seven stimuli below .03 indicates high internal consistency and since the calculated AD value in this study was .02, it indicated high internal consistency and that the observed proportion agreed with the theoretical proportions.

v. *Chi-square χ^2 test of significance for the paired comparison*

The purpose of the test is to determine whether or not the assumptions of normality of distribution of the discriminal processes, unidimensionality of the psychological continuum and equality of various values of standard deviations of the differences were met.

The first step was to convert the empirical proportions (P) and the theoretical proportions (P′) obtained for seven segments in Table 3.8 and 3.12 into corresponding Q values with the help of conversion Table II (Edwards, 1958). Tables 3.14 and 3.15, respectively, give the Q matrix.

TABLE 3.14 : Q values corresponding to the empirical proportions (P)

Segment	7	5	6	2	4	3	1
7	—						
5	41.55	—					
6	29.33	33.21	—				
2	27.97	31.95	43.85	—			
4	21.97	26.56	36.87	35.67	—		
3	16.43	20.27	31.95	33.21	40.40	—	
1	18.44	18.44	30.66	29.33	39.23	33.21	—

TABLE 3.15 : Q′ values corresponding to the theoretical proportions (P′)

Segment	7	5	6	2	4	3	1
7	—						
5	41.27	—					
6	29.20	32.65	—				
2	28.73	32.14	44.43	—			
4	22.79	25.92	37.64	38.17	—		
3	18.91	21.81	32.90	33.49	40.11	—	
1	15.89	18.53	28.86	29.40	35.85	40.69	—

The discrepancy of each value i.e. Q – Q′ is given in Table 3.16.

TABLE 3.16 : Q – Q′ values

Segment	7	5	6	2	4	3	1
7	—						
5	0.28	—					
6	0.13	0.56	—				
2	–0.76	–0.19	–0.57	—			
4	–0.82	–0.64	–0.77	–2.50	—		
3	–2.48	–1.54	–0.95	–0.19	0.29	—	
1	2.55	–0.09	1.80	–0.07	3.38	–7.48	—
$\Sigma(Q - Q')^2$	13.998	3.139	5.060	6.291	11.508	55.950	

The following formula for test of significance (Edwards, 1958) was used:

$$\chi^2 = \frac{\Sigma(Q - Q')^2}{821/N} = \frac{95.946}{16.42} = 5.843$$

where,

N = number of judges = 50

$$d.f. = \frac{(n-1)(n-2)}{2} = \frac{(7-1)(7-2)}{2} = 15$$

where,

N = number of segments = 7

The calculated χ^2 value 5.843 at 15 *d.f.* is lower than the table value of 24.996 at 0.05 level of significance. This indicates that the assumptions involved in finding the scale values of the given segments are tenable.

The internal consistency check with average discrepancy value and χ^2 test of significance showing that the assumptions are tenable, make the paired comparison scale a valid measure of rural farm women's role performance *vis-a-vis* the relative importance of the seven studied role segments of cattle and buffalo production. Content validity was insured in the collection, grouping and selection of roles for the construction of this scale.

vi. *Reliability of the scale*

Reliability of the paired comparison scale was found out by testing the inconsistency in judgements of the judges and the agreement among the judges.

In making paired comparison judgements, a subject may sometimes be inconsistent. An inconsistency in judgements occurs whenever there is a circular triad present in the *n(n-1)/2* judgements. The greater the number of circular triads occurring in the set of *n(n-1)/2* comparative judgements of a given subject, the more inconsistent the subject may be said to be and *vice versa*. The comparative judgements for a judge with a number of circular triads, is given in Table 3.17.

TABLE 3.17: Comparative judgements for a judge with a number of circular triads

Segment	7	5	6	2	4	3	1
7	—	1	1	1	1	1	1
5	0	—	0	1	1	1	1
6	0	1	—	0	1	1	1
2	0	0	1	—	0	0	1
4	0	0	0	1	—	0	1
3	0	0	0	1	1	—	
1	0	0	0	0	0	0	—
a	0	2	2	4	4	3	6
a^2		4	4	16	16	9	36

Circular triads were found out for each individual judge by the coefficient of consistence (Edwards, 1958):

$$\zeta = 1 - \frac{24d}{n(n^2-1)}$$

where

$d = \left(\frac{1}{12}\right) (n) \quad (n-1) \quad (2n-1)\frac{1}{2} \quad \times \quad \Sigma a^2$

n = number of segments

a = number of times segments '*i*' (column stimulus is chosen over '*j*' (row stimulus)

Therefore,

$$d = \left(\frac{1}{12}\right) 7 \times 6 \times 13 - \frac{1}{2} \times 85$$

$$= 45.50 - 42.50$$

$$= 3$$

and,

$$\zeta = 1 - \frac{24d}{n(n^2 - 1)}$$

$$= 1 - \frac{24 \times 3}{7(7^2 - 1)}$$

$$= 1 - \frac{72}{336}$$

$$= 0.786$$

The following formula (Edwards, 1958) was used to check the significance of coefficient of consistence:

$$\chi^2 = \left(\frac{8}{n-4}\right) \left(\frac{1}{4} nC_3 - d + \frac{1}{2}\right) + d.f.$$

where

n = number of stimuli

nC_3 = Number of combinations of n things taken 3 at a time or $n!/3!\,(n-3)!$

d = Observed number of circular triads

$$d.f. = \frac{n(n-1)\quad(n-2)}{(n-4)^2} = 23.33$$

Therefore,

$$\chi^2 = \left(\frac{8}{7-4}\right) \left(\frac{1}{4} \times 35 - 3 + \frac{1}{2}\right) + 23.33$$

$$= (2.66) \quad (0.25 \times 35 - 3 + 0.5) + 23.33$$

$$= 39.955$$

The table value of χ^2 at 0.01 probability level was found to be 41.638 (*df*, 23) which is greater than calculated value of 39.955 making it non-significant. Thus we may conclude that judges have been consistent in their judgements.

vii. *Coefficient of Agreement*

To found out the extent to which 50 judges agreed in their comparative judgements, the coefficient of agreement was calculated by using Kendall's test of significance '*u*', using the following formula (Edward, 1958):

$$u = \frac{2\tau}{(mC_2) \quad (nC_2)} - 1$$

Where,

m = number of judges

n = number of segments

$$mC_2 = \frac{m(m-1)}{2}$$

$$nC_2 = \frac{n(n^2-1)}{2}$$

$$\tau = (\Sigma fij^2 - m\,\Sigma fij) + (mC_2)\,(nC_2)$$

and where,

Σfij = sum of *fij* entries below the diagonal

The frequency with which the column segments were judged more favourably than the row segments by 50 judges is indicated in Table 3.18.

TABLE 3.18 : Frequency with which the column segment was judged more favourable with the row segment

Segment	7	5	6	2	4	3	1
7	—						
5	22	—					
6	12	15	—				
2	11	14	24	—			
4	12	10	18	17	—		
3	4	6	14	15	21	—	
1	5	5	13	12	20	15	—
Σfij	66	50	69	44	41	15	
Σfij^2	934	582	1265	658	841	225	

$$\Sigma fij = 285$$
$$\Sigma fij^2 = 4505$$
$$m = 50$$
$$n = 7$$

$$mC_2 = \frac{m(m-)}{2} = \frac{50(50-)}{2} = 1225$$

$$nC_2 = \frac{n(n-1)}{2} = \frac{7(7-1)}{2} = 21$$

$$\tau = (4505 - 50 \times 285) + (1225)(21)$$
$$= 4505 - 14250 + 25725$$
$$= 15980$$

Then,

$$u = \frac{2\tau}{(mC_2)(nC_2)} - 1$$
$$= \frac{2 \times 15980}{25725}$$
$$= 1.242 - 1$$
$$= 0.242$$

The minimum expected value of u for even number of judges was calculated using the following formula (Edwards, 1958).

$$u = \frac{-1}{(m-1)} = \frac{-1}{49} = -.020$$

As the calculated value of u (0.242) is positive and greater than the minimum expected value of 'u', it indicated agreement among the judges in their comparative judgement.

viii. χ^2 *Test for the Coefficient of Agreement*

Kendall's test of significance for u, based upon the χ^2 distribution was applied using the following formula, as 'm' was greater than 6 and 'n' was greater than 4:

$$\chi^2 = \left[\frac{4}{m-2}\right]\left[T - \frac{1}{2}(nC_2)(mC_2)\left(\frac{m-3}{m-2}\right)\right]$$

$$= \left[\frac{4}{50-2}\right]\left[15980 - \frac{1}{2}(21)(1225)\frac{50-3}{50-2}\right]$$

$$= [0.083] \quad [15980 - 12592.38]$$

$$= 281.17$$

$$d.f. = nC_2 \frac{m(m-1)}{(m-2)^2}$$

$$= 21\frac{50\,(5033-1)}{(50-2)^2}$$

$$= 21 \times \frac{2450}{2304}$$

$$= 22.33 \text{ or } 22$$

The calculated Kendall's χ^2 value i.e. 281.17 is greater than table value (40.289) at 1 per cent level of significance ($df = 22$) indicating that there was a high degree of agreement among the judges in their comparative judgements and the difference (u = 0.242) was only due to chance.

3.5 **Tools of Data Collection**

Data were collected from the selected respondents by using the questionnaire and interview schedule developed for the purpose. The required information for schedule was collected by going through relevant literature on the subject, discussions with experts and keeping in view the objectives of the study.

3.5.1 *Pre-testing*

Before giving a final shape to the interview schedule, it was pre-tested in an almost identical village situation with 30 non-respondent farm women having livestock. On the basis of experience in pretesting and analysis of the data so collected appropriate changes in the format and the content of the instrument were made. Data were then collected through so modified interview schedule.

3.5.2 *Data Collection*

The data were collected through personal interviews during May-June-July, 1988.

3.6 **Statistical Methods Used**

After collection, the data were compiled, tabulated and analysed keeping in view the objectives of the study. The following statistical methods were used in analysing the data using the standard formulae:

i. *Frequency* – It was used to find out the number of respondents in a particular cell.
ii. *Mean* – Mean was calculated by adding the total scores obtained by the respondents and dividing it by the total number of practices.
iii. *Percentage* – This was used for making simple comparisons. For calculating the percentage, the frequency of a particular cell was divided by the total number of respondents in that particular category and multiplied by 100.
iv. *Correlation Coefficient* – It was used to find out the relationship between the independent and dependent variables.
v. *Multiple regression analysis* – The multiple regression analysis

was used to determine the degree to which the dependent variables could be predicted with the help of independent variables.

a. *Multiple Regression*

The following multiple regression equation was used in this study :

$$Y_{1j} = \beta_0 + \beta_1 x_{1j} + \beta_2 x_{2j} + - \beta_i x_{ij} + \beta_k x_{kj} + e_{ij} \quad (1)$$

where,

Y_{1j} = the value of first dependent variable.
β_0 = the intercept made by the plane on Y axis i.e. a constant.
β_1 = the partial regression coefficient associated with variable X_i considering the remaining independent variable as fixed.
k = number of independent variables.
X_{ij} = value of ith independent variable of jth household.
e_{ij} = Error term associated with ijth observation and is assumed to be independent $N\sim(0, \sigma e_1^2)$

$$Y_2 j = \beta_0 + \beta_1 x_{ij} + \beta_{2x2j} + \ldots \beta_{iXij} + \beta_{kXkj} + e_{2j} \quad (2)$$

where,

Y_{2j} = the value of second dependent variable.
β_0 = the intercept made by the plane on the Y-axis i.e. a constant.
β_i = the partial regression coefficient associated with variable X_i and represents the amount of expected change in Y for per unit change in x_i considering the remaining independent variables as fixed.
k = number of independent variables.
x_{ij} = value of ith independent variable of jth individual.
e_{2j} = Error term associated with ij^{th} observation and is assumed to be independent $N\sim(0,\sigma e_2^2)$

b. *Multiple Correlation*

The explanatory predictive power of the multiple regression equation was evaluated with the help of multiple correlation coefficient R and the square of the multiple correlation coefficient R^2 (the coefficient of determination). The square of multiple correlation coefficient R^2 represents the extent of variation in independent variable as explained by all the independent variables taken together in multiple regression equation.

The significance of R^2 (coefficient of determination) was tested with the help of F test and the formula used was as follows:

$$F = \frac{R^2}{1 - R^2} \times \frac{N - K - 1}{K}$$

where,

F = variance ratio
K = number of independent variables
N = number of farm women in the sample

(vi) *Analysis of variance* – The significant differences between the effects of levels of any factor were tested by analysis of variance (ANOVA) technique. The critical difference test was also used for pairwise comparison between the means and the calculated value of '*F*', which was significant at decided level of significance.

(vii) *Path coefficient analysis* – Path coefficient analysis was done to determine the extent of contribution of each independent variable directly to the dependent variable and indirectly through dominant one.

(viii) *Paired 't' test* – The significance of mean of differences of the responses in relation to role expectation, perception and performance were tested by paired 't' test.

4

RESULTS AND DISCUSSION

This chapter deals with the profile of the respondents, the involvement of farm women in decision-making, role expectation of spouses, role perception and performance of farm women, differences between role expectation of spouses and role perception and performance of farm women according to different categories of socio-economic status, zero order correlation analysis of independent variables with dependent variables, multiple regression analysis of dependent variables with independent variables and direct and indirect effects of selected independent variables on dependent variables.

4.1 **Profile of the Respondents**

The study was conducted on a sample of 120 farm women, owning milch animals and living with her spouse under the same roof, drawn randomly from 12 villages in 4 blocks of Bareilly district.

The personal characteristics of the respondents affect their role in a given family. It was therefore, felt necessary to study their various characteristics such as age, caste, education etc. The following section discusses the profile of the respondents, in general, and characteristics of the farm women, in particular.

4.1.1 *Age and Family Position*

Table 4.1 points out that majority (55.83%) of the respondents were housewives while mothers-in-law and daughters-in-law almost equal in number (about 22%).

Majority (58.34%) of them belonged to the middle age group while about 33 per cent fell in the old age group. Only about 8 per cent were below 30 years of age.

TABLE 4.1: Distribution of respondents according to age and family position

Family position / Age group	Housewife f	Housewife %	Daughter-in-law f	Daughter-in-law %	Mother-in-law f	Mother-in-law %	Total f	Total %
Young 30 years	2	1.66	8	6.67	—	—	10	8.33
Middle 30-40 years	50	41.67	18	15.00	2	1.67	70	58.34
Old	15	12.50	1	0.83	24	20.00	40	33.33
Total	67	55.83	27	22.50	26	21.67	120	100.00

Table 4.1 further indicates that among housewives, majority of them belonged to middle age group whereas only two were below 30 years of age. Majority of daughters-in-law also belonged to middle age group, though over 6 per cent of them were below 30 years of age and only one respondent was above 40 years. On the other hand, all mothers-in-law were from the old age group except 2 who were from the middle age group.

4.1.2 *Caste*

Traditionally the village social system has been divided into various castes. The respondents were grouped into six major caste categories. Murao, Lodhe, Kissan and Taili were categorised under low caste. Carpenters, tailors and blacksmiths were placed under artisan caste. Prestige castes were, Thakurs, Goldsmiths etc. while Muslims were categorised under agriculture caste. Brahmins were categorised under dominant caste.

TABLE 4.2: Distribution of respondents in relation to caste

Caste Category	Frequency	Per cent
Scheduled caste	8	6.67
Lower caste	55	45.83
Artisan caste	2	1.67
Agricultural caste	32	26.67
Prestige caste	20	16.66
Dominant caste	3	2.50
Total	120	100.00

Table 4.2 reveals that majority (45.83%) of the respondents belonged to lower caste, followed by agriculture caste (26.67%). Over 16 per cent belonged to different prestige castes. Scheduled castes were only 6.67 per cent, while very few respondents were of dominant caste (2.50%) and artisan caste (1.67%).

4.1.3 *Occupation*

In the study area, majority of the respondents (80.94%) had farming as their major occupation on which their livelihood depended. Service and labour provided occupational livelihood to 14 respondent families. Other occupations such as caste occupation, business and independent profession, provided livelihood only to about 8 per cent of respondents' families (Table 4.3).

TABLE 4.3: Distribution of respondents according to occupation

Occupation	Frequency	Per cent
Labourer	7	5.83
Caste occupation	4	3.33
Business	3	2.50
Indep. profession	2	1.67
Agriculture	97	80.84
Service	7	5.83
Total	120	100.00

In most of the cases, the family had more than one occupation for their source of income. The earnings from all sources of income were, however, pooled in the family. The data in Table 4.3 show the major occupation on which the family depended for its livelihood.

4.1.4 *Education*

The study area (Bareilly Distt.) has very low literacy rate, particularly in case of rural farm women. Ninety per cent of the randomly selected farm women were illiterate while only 10 per cent of them were found literate.

TABLE 4.4: Distribution of respondents according to their educational status

Education	Frequency	Per cent
Illiterate	108	90.00
Can read only	—	—
Can read and write	2	1.67
Primary	8	6.67
Middle	1	0.83
High School	1	0.83
Total	120	100.00

Table 4.4 reveals that among literate farm women only two could read and write and only 8 were educated up to primary level. No farm women was educated above high school. The low level of education among farm women indicates traditional backwardness on one hand and the lack of infrastructural facilities for girls' education on the other, making them inaccessible to the newspapers and other print media and restricting them to limited channels of communication.

4.1.5 *Type of Houses*

The type of house is an important indicator of the economic as well as the social status of its dweller. Over 68 per cent of houses were either *kuccha* or mixed type while about 21 per cent respondents were having *pucca* houses. Only 3 respondents had thatched huts and 10 had mansions to live in.

TABLE 4.5: Distribution of respondents according to type of houses

Type of house	Frequency	Per cent
No house	—	—
Hut	3	2.50
Kuccha	47	39.17
Mixed	35	29.17
Pucca	25	20.83
Mansion	10	8.33
Total	120	100.00

Table 4.5 shows the rising trend of standard of living in the rural

area. The time utilization pattern of farm women varies on the basis of type of house in which she lives. They are generally responsible for cleaning, sweeping, maintenance and upkeep of houses. *Pucca* house saves a lot of time and energy from mud plastering and other maintenance activities which need to be done regularly in *kuccha* dwellings.

4.1.6 *Type and Size of Family*

The role of women in a rural family largely depends on the type and size of family. The jobs and responsibilities are well distributed among women in a joint family where more than one woman is available for the distribution of work, whereas in a nuclear family the housewife has to perform all the work alone. It is generally said that the joint family system is being replaced with nuclear family system with the development of technology and increasing mobility for career advancement by the educated youth. In case of the selected respondents, however, 70.83 per cent belonged to joint families.

TABLE 4.6: Distribution of respondents according to family type and family size

Family size \ Family type	Type of family					
	Nuclear		Joint		Total	
	f	%	f	%	f	%
5 and below family members	20	16.67	27	22.50	47	39.17
Above 5 family members	15	12.50	58	48.33	73	60.83
Total	35	29.17	85	70.83	120	100.00

Table 4.6 shows that more than 60 per cent of the respondent families had more than 5 family members. The table further reveals that only 12.50 per cent of the nuclear families had a family size of more than 5 whereas over 48 per cent of joint families had this size of family. Only 22.50 per cent of joint families and 16.67 per cent of nuclear families had up to 5 members in the family. The role of farm women in large families is more specific. They share more burden in respect of dairy farming operations. The data fulfils the expectation

that the joint families are large sized while nuclear are smaller.

4.1.7 *Dairy Herd Size and Size of Land-holding*

The number of milch animals possessed by a family influences the quantitative and qualitative aspects of role and responsibilities of the farm women.

Table 4.7 also reveals that 15 per cent of the farm families did not own any land. About 49 per cent of farm families possessed less than 1 hectare of land and belonged to the marginal farmers' category. Over 21 per cent of the farm families had 1-2 hectare of land and fell in the small farmers' category. A little over 14 per cent of the farm families possessed more than 2 hectare of land and belonged to medium (9.60%) and large (5.00%) farmers' categories.

TABLE 4.7: Distribution of respondents according to size of land-holding and dairy herd size

	Land-holding					
No. of milch animals	No land	1 ha	1-2 ha	2-4 ha	4 ha	Total
1-2	8 (6.67)	36 (30.00)	17 (14.17)	3 (2.50)	1 (0.83)	65 (54.17)
3-4	8 (6.67)	18 (15.00)	6 (5.00)	6 (5.00)	—	38 (31.66)
5-6	1 (0.83)	4 (3.33)	3 (2.50)	1 (0.83)	3 (2.50)	12 (10.00)
6	1 (0.83)	1 (0.83)	—	1 (0.83)	2 (1.67)	5 (4.17)
Total	18 (15.00)	59 (49.17)	26 (21.67)	11 (9.16)	6 (5.00)	120 (100.00)

Note: Figures in parentheses indicate percentage.

Table 4.7 further reveals that 8 landless, 36 marginal, 17 small, 3 medium and only 1 large farm families had less than 3 dairy animals. Nine landless, 22 marginal, 9 small, 7 medium and 3 large farm families had more than 3 and up to 6 milch animals. Only 1 landless, 1 marginal, 1 medium and 2 large farm families possessed more than 6 milch animals. The data also reveal that more than 54 per cent of the farm families possessed one or two milch animals, about 32 per cent kept 3-4 milch animals and only about 14 per cent had more than 4 milch animals.

4.1.8 *Green Fodder Cultivation*

Table 4.8 points out that majority (49.17%) of the respondents were possessing less than 1 hectare of land, while 21.66 per cent were found to have 1-2 hectare of land. Very few (5.00%) were possessing more than 4 hectare of land.

TABLE 4.8: Distribution of respondents according to size of land holding and area under fodder cultivation

Fodder cultivation	Size of land-holding					
	No land	1 ha	1-2 ha	2-4 ha	4 ha	Total
No fodder cultivation	18 (15.00)	35 (29.17)	15 (12.50)	1 (0.83)	—	69 (57.50)
0.1 - .3 ha	—	19 (15.83)	7 (5.83)	6 (5.00)	1 (0.83)	33 (27.50)
0.4 - 0.6 ha	—	3 (2.50)	4 (3.33)	1 (0.83)	2 (1.67)	10 (8.33)
0.7 - 0.9 ha	—	2 (1.67)	—	3 (2.50)	3 (2.50)	8 (6.67)
Total	18 (15.00)	59 (49.17)	26 (21.67)	11 (9.16)	6 (5.00)	120 (100.00)

Note: Figures in parentheses indicate percentage.

Table 4.8 further reveals that majority of the respondents (57.50%) were not cultivating green fodder, while 27.50 per cent of the respondents were found to cultivate green fodder in 0.1-0.3 hectare area. Over 8 per cent of the respondents were growing green fodder in an area of 0.4-0.6 hectare. A limited number of respondents were growing green fodder in an area of 0.7-0.9 hectare of land. It can thus be observed that since majority of the respondents were belonging to marginal and small farm families, hence food production for their families was the primary need. It is not always possible for such farm families to spare land for green fodder cultivation for their dairy animals. In spite of this limitation, it can be observed that out of the 51 respondents who were growing green fodder, 26 respondents belonged to marginal and small farmers' categories and were growing green fodder on an area of 0.1 to 0.3 hectares of their land and 9 were growing on more than 0.3 hectares. Area under fodder cultivation is, obviously dependent on the size of land-holding of the farmer and is secondary to the requirement of the food for the family of the livestock owner.

4.1.9 *Material Possessions*

The data presented in Table 4.9 indicate that about 70 per cent of the respondent families owned bicycles. Less than one third of the respondents owned radio and about 9 per cent owned television which indicates that the important mass media sources are owned by very limited number of people in the study area.

TABLE 4.9: Distribution of respondents according to material possessions

Material possession	Frequency	Per cent
Bullock cart	15	12.50
Bicycle	83	69.17
Radio	36	30.00
Chairs	18	15.00
Tube well	16	13.33
Pumping set	24	20.00
TV	11	9.17
Improved agril. implements	6	5.00

Table 4.9 further pointed out that one-third of the respondent families owned pumping sets/tubewells for the irrigation of their fields. They probably not only irrigate their own fields but also provide water to others' fields.

Other materials like bullock cart, chairs and improved agricultural implements were found to be possessed by 12.50, 15.00 and 5.00 per cent of respondents, respectively. It can be observed that the respondents in general had rather limited material possessions which indicates low economic status among the studied farm women.

4.1.10 *Socio-economic Status*

It was intended to classify the respondents according to socio-economic status into different socio-economic categories viz. lower, lower-middle, middle, upper-middle and upper. The socio-economic status scale of Trivedi (1963) was used to assign scores to each of the respondent. The selected respondents were then classified according to the scores obtained by them.

The data presented in Table 4.10 indicates that 85 per cent of the respondents belong to lower-middle or middle socio-economic status categories. Only about 13 per cent belonged to upper socio-economic strata and 1.67 per cent to the lower class. The socio-economic status of the family influences, to a considerable extent, the adoption of improved practices, both directly and indirectly. It may also affect the participation of farm women in animal husbandry activities.

TABLE 4.10: Distribution of respondents according to category of socio-economic status

Categories of socio-economic status	Frequency	Per cent
Lower (Below 10 scores)	2	1.67
Lower-middle (10-20 scores)	60	50.00
Middle (21-31 scores)	42	35.00
Upper-middle (32-42 scores)	11	9.17
Upper (43 scores & above)	5	4.16
Total	120	100.00

According to the prevailing norms, the involvement of farm women in animal husbandry tasks decreases with the increase in socio-economic status.

4.1.11 *Productivity of Dairy Animals*

Productivity for a selected family was measured on the basis of average milk production per lactation per dairy animal. Productivity index was then calculated which was expressed in terms of percentages (Yang, 1980). On the basis of productivity index, the respondents were grouped into three categories viz. low, medium and high productivity level.

The data presented in Table 4.11 indicates that the respondents having lower-middle and middle socio-economic status were found to have medium to high productivity level. The respondents belonging to lower socio-economic status (1.67%) had low levels of productivity.

TABLE 4.11: Distribution of respondents according to categories of socio-economic status and level of productivity

Socio-economic status	Productivity level							
	Below 43.80%		43.81-157.02%		157.03% & above			
	(Low)		(Medium)		(High)		Total	
	f	%	f	%	f	%	f	%
Lower (Below 10 scores)	2	1.67	—	—	—	—	2	1.67
Lower-middle (10-20 scores)	9	7.50	48	40.00	4	3.33	61	50.83
Middle (21-31 scores)	1	0.83	33	27.50	8	6.67	42	35.00
Upper-middle (32-42 scores)	—	—	10	8.33	1	0.83	11	9.17
Upper (43 scores & above)	—	—	3	2.50	1	0.83	4	3.33
Total	12	10.00	94	78.33	14	11.67	120	100.00

It is obvious from Table 4.11 that the level of productivity of the dairy animals increases with the increase in the socio-economic status of the family. This might be due to the reason that upper class livestock owners are able to divert more resources and devote more time to the care of dairy animals. They might be having the required motivation also to increase the productivity levels of their dairy animals.

4.1.12 *Family Education Status*

The knowledge, attitude and activities of an individual along with his/her role do not depend only on his/her own education but also on the overall educational level of the family because the educated members of a family may influence the overall outlook of other family members.

It was, therefore, considered appropriate to compute the family education status of the selected respondents. The respondents were then divided into three categories, viz. low or nil family education status, medium and high family education status, having scores from 0, up to 1.0 and more than 1.0 family education index, on the basis of means and standard deviation.

TABLE 4.12: Distribution of respondents according to family education status

Family education status	Score	Frequency	Per cent
Low	0	61	50.83
Medium	0.1 - 1.0	30	25.00
High	1.1 and above	29	24.17
Total		120	24.17

The data presented in Table 4.12 pointed out that over 50 per cent of the respondent families were illiterate or having zero family education status score. About 25 per cent of the respondents families were having medium family education status and about equal percentage were found to have high family education status. This shows that half of the respondents were illiterate while about one-fourth had a family education status up to primary level (medium family education status) and about the same percentage had a family educational status above the primary level (high family education status).

4.1.13 *Information Sources Utilization*

It was intended to study the various sources of information available with the farm women and their extent of utilization, as the level of their exposure to information sources influences their knowledge, attitudes, perceptions and performance of their roles. The respondents were asked to state the extent of utilization of three major categories of scores of information viz. non-institutional, institutional and mass media.

TABLE 4.13: Distribution of respondents according to information sources utilization

Sources/channels of information used	Degree of utilization											
	More than once a week		Once a week		Once a fort-night		Once a month		Once in more than a month		Never	
	f	%	f	%	f	%	f	%	f	%	f	%
Interpersonal												
A. *Non-institutionalized*												
i. Friends and relatives in village	1	0.83	1	0.83	5	4.17	11	9.17	47	39.17	7	5.83
ii. Friends and relatives outside the village	—	—	—	—	—	—	4	3.33	8	6.67	—	—
iii. Neighbours	1	0.83	1	0.83	4	3.33	35	29.17	35	29.17	—	—
B. *Institutionalized*												
i. V.L.W.	—	—	—	—	—	—	—	—	—	—	—	—
ii. Stockmen	—	—	—	—	—	—	—	—	7	5.83	—	—
iii. V.A.S.	2	1.67	—	—	1	0.83	5	4.17	21	17.50	—	—
iv. Dy. Director (AH)	—	—	—	—	—	—	—	—	—	—	—	—
Mass media												
i. Radio	—	—	2	1.67	—	—	—	—	12	10.00	—	—
ii. Dairy *mela*/ Kissan *mela*	—	—	—	—	—	—	—	—	8	6.67	—	—
iii. Television	4	3.33	3	2.50	—	—	—	—	—	—	—	—

The data in Table 4.13 pointed out that the majority of the farm women were found to utilize the non-institutionalized interpersonal sources within the village only. Majority of the respondents used non-institutionalized sources of information like friends and relatives and neighbours. Friends and relatives were the sources of information for 39.17 per cent of the respondents once in more than a month. A small proportion of respondents utilized this source more frequently. Neighbours were used as a source of information by 29.17 per cent of the respondents once in a month and the same percentage utilized this source once in more than a month.

Institutionalized sources like Veterinary Assistant Surgeon (17.50%) and Stockman (5.83%) were also utilized as information sources but once in more than a month.

Mass media were utilized by very limited number of respondents as sources of information. Radio was used by only about 10 per cent of farm women once in more than a month; whereas *Melas* were attended only by about 7 per cent of the respondents once in more than a month.

Table 4.13 further revealed that though the Television was utilized by only a very small percentage of farm women but this medium was used more frequently than any other source of information although by a small number of respondents. This indicates that TV was an attractive mass medium in the study area among the farm women and if made available to other respondents, this can serve as an effective medium of communication among the farm families.

4.1.14 *Knowledge Level of Farm Women*

The farm women are involved in decision-making process and performance of dairy animal production activities in varying degrees. The level of participation in decision making and how well a task is performed depends largely on the knowledge and skill levels possessed by the farm women about the given task. In view of this it was intended to study the knowledge and skill level of the farm women in various segments of dairy animal production.

The data on knowledge of farm women presented in Table 4.14 indicates that the majority farm women possessed low level of knowledge in breeding (97.50%), management (95.83%), marketing (95.00%), feeding (86.67%) and health care (80.00%). Majority of farm women (79.17%) had medium level of knowledge about milk products making while only 7.50 per cent of the respondents had high level of knowledge in this segment only. In the segment of green fodder production, the knowledge level was found to be low (61.67%) to medium (38.33%).

TABLE 4.14: Distribution of respondents according to level of knowledge in dairy animal production

Segment \ Score	Knowledge level							
	(0-29%)		(30-59%)		(60% & above)			
	Low		Medium		High		Total	
	f	%	f	%	f	%	f	%
Feeding	104	86.67	16	13.33	—	—	120	100.00
Breeding	117	97.50	3	2.50	—	—	120	100.00
Management	115	95.83	5	4.17	—	—	120	100.00
Health Care	96	80.00	24	20.00	—	—	120	100.00
Milk Products Making	16	13.33	95	79.17	9	7.50	120	100.00
Green Fodder Production	74	61.67	46	38.33	—	—	120	100.00
Marketing	114	95.00	8	5.00	—	—	120	100.00

The data suggested that level of knowledge of the farm women, in general, is low in highly important areas of feeding, management, breeding and health care, where they are involved more but lack in knowledge in these segments. Dube and Singh (1976), Sohal and Singh (1977), Rath (1978), Sohal and Tyagi (1978) and Pawar (1979) had also observed that the farmers had low level of knowledge in almost all the aspects of animal production.

4.1.15 *Skill Level of Farm Women*

The efficiency of work depends to a large extent on the level of skill possessed by the incumbent. Certain activities/tasks require more and complex skills than the others. In feeding, the activities like giving feed and fodder to animals, giving water to animals, chaffing the fodder and taking the animals for grazing, require less skill than the activity like preparing home made concentrate mixture. Similarly, some managemental activities, like milking etc., require more skill. Keeping this in view two such activities in which the farm women are usually involved, viz. clean milk production and preparation and feeding of home made concentrate mixture were selected to measure their levels of skill in these skilled activities, through skill tests developed for the study.

TABLE 4.15: Distribution of respondents according to level of skill

Score / Activities/Tasks	Skill level							
	(0-29%)		(30-59%)		(60% & above)			
	Low		Medium		High		Total	
	f	%	f	%	f	%	f	%
Clean milk production	80	66.67	40	33.33	–	–	120	100.00
Preparing and feeding home-made concentrate mixture	114	95.00	6	5.00	–	–	120	100.00

The data in Table 4.15 indicate that about 67 per cent of the respondents had low skill level in clean milk production, the remaining 33 per cent had medium skill level. No respondent had high skill level in this important activity.

In case of the other skill studied viz. preparation and feeding of home made concentrate mixture, about 95 per cent respondents had low skill level. The data suggest that there is acute necessity of developing skills in farm women through skill oriented training with respect to those activities where they are directly and regularly involved.

The data also suggest that skill oriented tasks performed by farm women need to be paid more attention. They need to be identified and their levels of performance studied. It is all the more important in production oriented skilled activities.

4.2 Involvement of Farm Women in Decision-Making

Contribution of farm women in dairy animal production activities was studied with respect to their level of involvement in the decision-making process. Obviously, involvement in decision-making of an individual family member may not be of the same level for all the aspects of dairy animal production. In certain matters, the farm women might take independent decisions while in others they might not be involved at all, the decisions being taken by their spouses and/or other family members. The involvement of farm women in decision-making was studied on each role segment separately viz. feeding, breeding, management, health care, milk products making, green fodder production and marketing. The farm women were asked to indicate who was taking decision in each activity of each of the seven segments on a four point continuum viz. decision taken by spouse alone (independent decision), spouse + housewife + other family members/ friends/ relatives (collective decision), housewife + spouse (joint decision) and housewife alone (independent decision). In the event of the respondent not responding to an activity, no response was recorded. As can be seen the continuum was in ascending order with respect to the perceived level of involvement in decision-making by the farm women, the spouse being an independent decision maker at one end of the continuum and the farm women being the independent decision maker at the other, with the involvement of the farm women gradually increasing from one end of the continuum to the other.

4.2.1 *Feeding*

Feeding activities of dairy animals are generally performed by farm women except taking the animals for grazing and chaffing the fodder although they usually are involved in the performance of these activities also. Obviously, the individual who performs the activities most of the time should also be involved in the decision-making process in those matters.

The data presented in Table 4.16 revealed that the farm women were involved in the decision-making process in most of the activities either independently or jointly with the spouse. It can be seen that in matters like whether or not concentrates are to be fed (45.83%), what concentrates are to be fed (57.50%), how much concentrates are to be fed (79.17%), how much crop residues are to be fed (60.00%), most of the farm women were reported to be taking decisions independently. In matters like whether concentrates are to be homemade or purchased, whether or not the green fodder has to be fed and whether or not the green fodder is to be chaffed, in most cases the decision was taken either by the housewife alone or jointly by the housewife and the spouse. In matters like whether the green fodder was to be purchased or home grown (41.67%) and how much green fodder was to be fed (44.17%), most of the time the decisions were taken jointly by the housewife and the spouse. In the case of how the straws were to be stored, the spouse took the decision, most of the time (44.16%).

TABLE 4.16: Involvement of farm women in decision-making — FEEDING

(*N = 120*)

Sl. No.	Activities/Tasks	Decision-making pattern				No response
		Spouse only (%)	Collective (%)	Joint (%)	Farm woman only (%)	(%)
1.	Concentrates to be fed or not	16.67	10.00	27.50	45.83	—
2.	What concentrates are to be fed	21.67	7.50	13.33	57.50	—
3.	Quantity of concentrates are to be fed	10.83	2.50	7.50	79.17	—
4.	Concentrate mixture to be home made or purchased	35.83	5.00	31.67	27.50	—
5.	Mineral mixture to be fed or not	42.50	8.33	15.83	31.67	1.67
6.	Frequency of feeding of minerals	40.00	10.00	10.83	37.50	1.67
7.	Green fodder to be fed or not	24.17	10.00	30.83	35.00	—
8.	Green fodder to be purchased or home made	35.83	13.43	41.67	5.00	4.07
9.	Quantity of green fodder to be fed	23.23	14.26	44.17	16.67	1.67
10.	Quantity of crop residues are to be given	11.67	3.33	22.50	60.00	2.50
11.	Silage or hay making	4.17	2.50	5.00	21.67	66.66
12.	Chaffing of fodder	26.67	21.66	31.67	18.33	1.67
13.	Method of storage of straws	44.16	24.17	5.83	24.17	1.67
	Overall average	25.95	10.21	22.18	35.39	6.27

Collective = All family members/relatives/friends.
Joint = Spouse and farm woman only.

It would be seen that in matters involving monetary input, most of the decisions were taken jointly and the farm women were mostly involved in taking decisions about the activities which did not have monetary inputs.

Table 4.16 also indicates that silage/hay making is not practised by majority of the farmers and livestock owners in the study area, and therefore, a high percentage of respondents (66.66%) did not respond to this activity.

The findings of this study are in a consonance with the results of Dubey, *et al.* (1982) that housewife is involved in decision-making process in feeding the dairy cattle and buffaloes. Verma (1984) had also observed that decision with regard to feeding were taken either jointly or by housewife alone.

4.2.2 *Breeding*

Table 4.17 indicate that farm women are not much involved in the decision making process with regard to breeding activities of dairy animals, whether it is a decision about going for artificial insemination or for natural service or whether or not to get the pregnancy confirmed through pregnancy diagnosis. Dubey, *et al.* (1982) had also reported similar results.

Probably the reason why the decision in breeding are mostly taken by farmers, and not by farm women, is that most of the breeding activities are outdoor activities requiring the animal to be taken outside the home to the veterinary hospital which is generally located at a distance from the village. This is time consuming and the housewife with her other domestic and livestock related activities does not perceive herself to be able to spare that much of time to attend to these outdoor and time consuming activities. The decision, therefore, is taken by the spouse (livestock owner) who may be able to attend to other jobs also while performing these outdoor activities.

4.2.3 *Management*

Management is a very important segment of dairy animal production which, by and large, has indoor activities but many of these require monetary inputs. Table 4.18 shows that decisions about weaning of calf (53.33%), keeping the animal in the open or in shed (57.50), time of milking (80%) and number of times the animal is to be milked

(80%) were taken by housewife alone whereas decisions about whether the shed is to be *pucca* or *kutcha* was most of the time decided jointly by the farm women and the spouse (42.50%), although 30 per cent of the respondents also reported that the decision was taken by the spouse alone. It will be further observed that the number of dairy animals which were to be kept by the family was mainly a joint decision of the housewife and the spouse although about 27 per cent respondents also reported that it was a collective family decision. As regards the disposal of dead animals, 37.50 per cent of respondents reported that the spouses alone decided this matter but more than 33 per cent said that it was a collective decision. It is, however, obvious that the spouse is the main decision-maker in this matter probably because it is an outdoor activity and involves physical labour. More than 18 per cent of the respondents had, however, not responded to the question.

TABLE 4.17: Involvement of farm women in decision-making — BREEDING (*N = 120*)

Sl. No.	Activities/Tasks	Decision making pattern				No response
		Spouse only (%)	Collec-tive (%)	Joint (%)	Farm woman only (%)	(%)
1.	A.I. or natural service	82.50	8.33	7.50	–	1.67
2.	Selection of site in case of natural service	83.33	10.00	6.67	–	–
3.	Treatment of animals suffering with reproductive disorders to be got done by Vety. doctor or village healer	78.34	13.33	7.50	0.83	–
4.	Pregnancy diagnosis to be got done or not	70.00	8.33	10.84	–	10.83
	Overall average	78.54	10.00	8.13	0.21	3.12

TABLE 4.18: Involvement of farm women in decision-making — MANAGEMENT (*N = 120*)

Sl. No.	Activities/Tasks	Decision-making pattern				No response
		Spouse only	Collec-tive	Joint	Farm woman only	
		(%)	(%)	(%)	(%)	(%)
1.	Weaning of calf to be done or not	4.17	8.33	6.67	53.33	27.50
2.	Animals to be kept in open or in shed	2.50	21.67	17.50	57.50	0.83
3.	Shed to be *pucca* or *kuccha*	30.00	12.50	42.50	15.00	—
4.	Colostrum to be fed to the newborn calf or not	6.67	20.83	—	—	72.50
5.	Number of dairy animals to be kept	20.00	26.67	41.66	11.67	—
6.	Disposing off dead animals	37.50	33.43	7.50	3.33	18.33
7.	Shed to be disinfected or not	25.00	20.00	5.00	2.50	47.50
8.	Number of times animals is to be milked	—	12.50	7.50	80.00	—
9.	Time of milking	2.50	13.33	4.17	80.00	—
10.	Number of times animal is to be bathed and groomed in a week	26.66	11.67	5.00	25.00	31.67
	Overall average	15.50	18.09	13.75	32.83	19.83

It would be observed that in this segment also activities requiring monetary input were decided either jointly by the farm women and the spouse or collectively by the family. Activities like feeding of colostrum and whether or not the shed is to be disinfected were not responded to by a large number of respondents, probably because either they did not perceive that any decision was required to be taken for these matters as they were performed as a routine or they are not aware of these important management practices and are, therefore, not performing these activities.

4.2.4 *Health Care*

It is interesting to note from Table 4.19 that a high percentage of respondents did not respond to the activity like whether or not to get the animals vaccinated (56.67%) and regularly dewormed (66.67%). This phenomenon may be because these very important health care activities are not being practised by the farm families in the area of study. However, 30 per cent of the respondents reported that the decision about vaccination to be got done or not was taken by the spouse only. About 14 per cent of the respondents, however, reported that the matter of regularly getting the animals dewormed was decided by the spouses only. A small percentage of respondents (18.33%) reported that this decision was taken jointly by the farm women and her spouse. Regarding the decision as to from whom should the sick animal be got treated, majority of the respondents reported that the decision was taken either by the spouse himself (40%) or it was decided in consultation with the farm woman (22.50%) or that it was a collective decision (26.67%).

TABLE 4.19: Involvement of farm women in decision-making — HEALTH CARE

(*N* = *120*)

Sl. No.	Activities/Tasks	Decision-making pattern				No response
		Spouse only (%)	Collective (%)	Joint (%)	Farm woman only (%)	(%)
1.	Vaccination to be got done or not	30.00	10.00	3.33	—	56.67
2.	Treatment of sick animals to be got done by vety. doctor or village healer	40.00	26.67	22.50	4.16	6.67
3.	Regular deworming to be got done or not	14.17	7.50	8.33	3.33	66.67
	Overall average	28.06	14.72	11.39	2.49	43.34

Dubey, *et al.* (1982) had, however, reported that the housewife was involved in decisions regarding the health care of dairy cattle and buffaloes.

4.2.5 *Milk Products Making*

The activities in the segment of milk products making, as is obvious from Table 4.20 are indoor activities and, therefore, fall strictly within the home domain of farm women. It would be seen from the table that in matters like how much milk was to be used at home for consumption (46.67%) and what product is to be made out of it (60.83%), is decided by the farm women herself although 20 per cent of respondents also reported that they were taking the decision about this activity in consultation with their spouse. About 29 per cent respondents said that it was a collective decision.

TABLE 4.20: Involvement of farm women in decision-making — MILK PRODUCTS MAKING

(*N = 120*)

Sl. No.	Activities/Tasks	Decision-making pattern				No response
		Spouse only (%)	Collec-tive (%)	Joint (%)	Farm woman only (%)	(%)
1.	Quantity of milk to be used for home consumption	3.33	29.17	20.00	46.67	0.83
2.	Surplus milk to be sold/processed for products	11.67	29.17	23.33	26.66	9.17
3.	Type of products to be made	7.50	26.67	5.00	60.83	—
	Overall average	7.50	28.34	16.11	44.72	3.33

The probable reason for this is that as a home-maker and housewife, she has to look and cater to the nutritional requirements, needs and tastes of the other family members, in these matters. In a matter like how much liquid milk or product should be converted into cash by selling it, about 27 per cent respondents reported taking independent decisions. More than half the respondents, however, reported that it is decided in consultation with her spouse (23.33%) or collectively (29.17%). The findings are in consonance with the conclusions arrived at by Puri (1972), Dubey, *et al.* (1982) and Saraswathi, *et al.* (1987).

4.2.6 *Green Fodder Production*

Fodder production is another segment which is an outdoor activity and requires monetary input by the family. As is obvious from Table 4.21 that in most of the families the decisions about all the activities in this segment were taken by the spouse of the respondent except for storage of fodder. Obviously, since fodder is to be stored at home, the farm women also has to give her input in its execution because she is associated with the decisions about the quantity of green fodder to be fed to the dairy animals (Table 4.16) and whether the green fodder is to be purchased or home grown. The decision, therefore, regarding this activity was taken either by the spouse himself (33.34%) or by the farm women alone (24.17%) or collectively (18.33%).

TABLE 4.21: Involvement of farm women in decision-making — GREEN FODDER PRODUCTION

(*N = 120*)

Sl. No.	Activities/Tasks	Decision-making pattern				No response
		Spouse only (%)	Collective (%)	Joint (%)	Farm woman only (%)	(%)
1.	Type of fodder to be grown	58.33	22.50	8.33	—	10.84
2.	Area in which fodder is to be grown	64.17	20.83	4.16	—	10.84
3.	Which seed should be used?	60.00	24.17	4.16	—	11.67
4.	Fertilizers to be used or not	60.00	21.67	4.16	—	14.17
5.	Number of Irrigations to be given	60.00	22.50	4.16	—	13.34
6.	Storage of fodder	33.34	18.33	5.83	24.17	18.33
	Overall average	55.97	21.66	5.14	4.03	13.20

TABLE 4.22: Involvement of farm women in decision-making — MARKETING (*N = 120*)

Sl. No.	Activities/Tasks	Decision-making pattern				No response
		Spouse only (%)	Collective (%)	Joint (%)	Farm woman only (%)	(%)
1.	Place of marketing the milk, if it is to be sold	36.66	45.00	6.67	—	11.67
2.	Quantity of surplus milk to be sold	23.34	30.83	10.00	20.83	15.00
3.	Products made to be sold or consumed at home	20.00	30.00	20.00	20.83	9.17
4.	Quantity of milk products to be sold	17.50	34.16	13.33	10.84	24.17
5.	Place of marketing the milk products	32.50	38.33	3.33	1.67	24.17
6.	Rate at which milk is to be sold	29.17	46.67	8.33	—	15.83
7.	Rate at which milk products are to be sold	30.83	43.33	6.67	—	19.17
8.	Place of purchasing concentrates, if required	40.83	50.00	9.17	—	—
9.	Place of purchasing green fodder, if required	38.33	50.00	11.67	—	—
10.	Purchase of farm/dairy equipments	40.83	50.00	2.50	—	6.67
11.	Number and type of animals to be sold	25.00	55.84	19.16	—	—
12.	Place of sale of animals	25.00	61.67	13.33	—	—
13.	Minimum price at which animal to be sold	22.50	68.33	9.17	—	—
14.	Type of animal to be purchased (crossbred, Improved or Desi)	20.83	66.67	12.50	—	—
15.	Place of purchase of animals	24.17	70.83	5.00	—	—
16.	Maximum price at which animal to be purchased	22.50	68.33	9.17	—	—
17.	Taking loan for purchase of animals	15.83	18.34	2.50	—	63.33
18.	Quantity of feed/green fodder to be purchased	24.17	27.50	15.00	—	33.33
	Overall average	27.23	47.54	9.86	3.01	12.36

4.2.7 *Marketing*

Table 4.22 reveals that in matters like place where milk and the milk products are to be sold or the place from where concentrates and fodder is to be purchased and the place where the animals are to be sold or purchased are decided by the spouse either independently or the family collectively. Farm women, however, do contribute when the matters like quantity of surplus milk to be sold and whether the product made at home is to be sold or consumed, are to be decided upon, as revealed in Table 4.22.

The farm women contribute in deciding these matters probably as a home-maker who has to look after the nutritional requirements of the family. In almost all other activities, the decision is taken by the spouse alone or in consultation with the family members although a small percentage of respondents did report that some activities like rate at which milk or product is to be sold, number and type of animals to be sold or purchased, place of purchasing concentrate or green fodder were decided jointly by her and the spouse.

Singh (1968), Sunita (1977), Dubey, *et al.* (1982), Kaur (1983) and Nand, *et al.* (1988) had reported that farm women are involved in joint decisions on marketing aspects. Puri (1972) on the other hand, had found that decision relating to marketing are taken by husbands. Seema (1986), Saraswathi, *et al.* (1987) and Ghosh and Adhikari (1988) had reported dominance of farm women in decisions relating to marketing. This study has, however, revealed that the level of involvement of farm women in marketing divisions is very low or nil except in matters where she has to cater to the nutritional requirements of the family as a home-maker, where it is a little higher than in other activities.

On the basis of the findings in Tables 4.16 to 4.22, profile of the decision-making pattern in rural households and the level of involvement of farm women in it, is presented in Table 4.23.

The significant points which emerged from the findings are as follows:

1. In more than one-third of the rural families, decisions regarding the overall activities associated with feeding of dairy animals were taken by farm women alone. Over 22 per cent of respondents, however, reported that the decisions in this segment were taken jointly by them and the livestock owner.

TABLE 4.23: Profile of decision-making pattern in dairy animal production in rural households

(*N = 120*)

Segment	Decision-making pattern*				No response
	Spouse alone (%)	Collective (%)	Joint (%)	Housewife alone (%)	(%)
Feeding	25.95	10.21	22.18	35.39	6.27
Breeding	78.54	10.00	8.13	0.21	3.12
Management	15.50	18.09	13.75	32.83	19.83
Health care	28.06	14.72	11.39	2.49	43.34
Milk products making	7.50	28.34	16.11	44.72	3.33
Green fodder production	55.97	21.66	5.14	4.03	13.20
Marketing	27.23	47.54	9.86	3.01	12.36

* Overall average scores.

2. In about one-third of the rural families, decisions regarding overall management of dairy animals were taken by farm women alone.
3. In about 45 per cent of rural families, farm women were taking decisions on the overall activities associated with milk products making. In about 16 per cent families, such decisions were taken jointly by the livestock owner and the farm woman. In more than one-fourth rural families decisions in this segment were, however, taken collectively.
4. In over three-fourths of the rural families, decision-making on the overall activities associated with breeding of the dairy animals was monopolised by the livestock owners. Only in about 8 per cent of rural families, joint decisions were taken by the livestock owner and the farm woman.
5. In more than half of the rural families, decisions on the overall activities associated with green fodder production were taken independently by livestock owners. In more than 21 per cent rural families decisions in this segment were, however, taken collectively.
6. In more than 47 per cent rural families, decisions on the overall marketing activities were taken collectively, but in more than

27 per cent families, livestock owners were taking independent decisions.

7. Although a clear decision-making pattern in the segment of health care of dairy animals did not emerge, because of a very large non-response but, on the basis of overall average, in 28 per cent rural families, decisions in this segment were taken by the livestock owners independently.
8. The findings indicate that, generally, the segments in which most of the activities are to be performed indoors, the farm women have a more dominant role to play in decision-making in rural families.
9. In segments where immediate or instant monetary inputs or outputs are involved, livestock owners tend to dominate the decision-making process.

4.3 Role Expectation, Perception and Performance in Dairy Animal Production

Role which a farm woman actually performs in a given situation is an interactive sum total of the expectations of the social system in which she lives, the institution of family in which she performs and her own perceptions of the tasks she is expected to perform. Her personal abilities, capabilities and competence also influence her performance of tasks substantially, both qualitatively as well as quantitatively.

It is obvious that the members of her family would have an immediate influence on her day to day performance and her spouse is expected to be the most influential person affecting her performance of the role which can be performed either harmoniously, if there is clear understanding of the role of the farm women, or in a conflicting situation in the absence of it, affecting the actual role performance both qualitatively and quantitatively. Studies have indicated that apart from the role of a home-maker, the farm women substantially share the agriculture and animal husbandry related tasks in our rural social system. It was, therefore, proposed to study the expectations of the spouse about the role of farm women in dairy animal production in different segments viz. feeding, breeding, management, health care, milk products making, green fodder production and marketing. Each role segment was sub-divided into various activities/tasks. The spouses of respective farm women, who had been automatically selected as respondents alongwith the randomly selected farm women, were asked to state their expectations as to who should perform those activities/ tasks,on a five point continuum. The responses were to interalia identify whether or not the farm woman was expected to perform the given activity in each segment.

An adult individual is likely to modify the performance of a role in the light of how she perceives it to be performed. The role performed by the farm women, therefore, partially depends upon her perceptions about it. The perception itself may, however, be influenced by her own interest, capabilities , abilities, knowledge, attitude etc. In order to study the perception of their role in dairy animal production, the respondents were asked to respond to each of the activities for which the data on the expectations of their spouses were collected, on a five point continuum.

The ultimate objective of this study was to analyse the role being actually performed by farm women in the dairy animal production *vis-a-vis* the expectations of their spouses and their own perceptions of each of the studied role segments. An individual farm woman's performance, however, depends, among many complex factors, on what is expected from her and how she herself perceives her role input in a given activity/task. Role performance of farm women was studied by asking the respondents, on a five-point continuum, as to how well they were performing each activity. They were asked to respond to the same activities on which responses on the expectations of the spouses and their own perceptions had been collected.

4.3.1 *Role Expectation : Feeding*

Table 4.24 points out that the expectations of the spouse from the farm women in the role segment of feeding was high (overall mean score, 3.48). This indicated that spouses expected from the farm women to perform most of the activities related to feeding of the dairy cattle and buffaloes. When the different activities/tasks related to the feeding of dairy animals, were studied, it was found that the spouses expected the farm women to perform at a higher level on the activities like 'give the feed and fodder to the animals', prepare concentrate mixture', 'give concentrate mixture to the animals', 'mix green fodder with roughage' and 'give drinking water to the animals' (Table 4.24, Fig. 4.1).

The spouses did not expect the farm women to perform high on the activities like 'taking the animals for grazing','chaffing the fodder' and 'giving mineral mixture/salt to the animals'.

The data presented in Table 4.24 indicate that the expectations of the spouses from the farm women were about their performing most of the indoor activities related to feeding while outdoor activities like taking the animals for grazing and those tasks which require physical labour like chaffing the fodder were not expected by them to be performed by the farm women. Cultural and social values and norms also appear to influence the expectations of the spouses from the farm women to the extent that the institution of family requires the farm women to stay home and attend to her other responsibilities as a home-maker.

4.3.2 *Role Perception : Feeding*

The perusal of Table 4.24 reveals that perception of farm women in the role segment of feeding was high (overall mean score, 3.16). They perceived this role segment as the one in which they could contribute at a high level. The role segment had been given the top ranking by the judges (paired comparison technique) in terms of its importance in cattle and buffalo production (Table 4.31). Table 4.24 and Fig. 4.1 further reveal that the activities of this role segment which were perceived to be required to be performed by her were 'giving the fodder to animals', 'giving feed to animals', 'mixing green fodder with roughage', 'giving concentrate mixture to animals', 'preparing concentrate mixture', and 'giving drinking water to animals'. It will be observed that all these activities are indoor activities requiring relatively less physical labour and energy.

4.3.3 *Role Performance : Feeding*

Table 4.24 further reveals that activities requiring outdoor performance and more physical labour like 'taking the animals for grazing' and 'chaffing the fodder' were not perceived by the farm women as a high role priority.

High performance was found among farm women in activities such as 'giving green fodder to animals', 'giving the feed to animals', 'preparing concentrates mixture', 'mixing green fodder with roughage', and 'giving concentrate mixture to animals'. All other activities/tasks on which data have been collected were found to be low in performance. The overall mean score for performance was found to be 3.32. The results are in consonance with the results reported by Dey (1977) and Gupta, *et al.* (1986).

TABLE 4.24: Mean scores of role expectation of spouses and role perception and performance of farm women — FEEDING

	Activities/Tasks	Expectation[a]	Perception[b]	Performance[c]
1.	Taking the animals for grazing	1.85	1.58	1.42
2.	Chaffing the fodder	2.38	2.11	2.05
3.	Mixing green fodder with roughage	3.92	3.71	3.94
4.	Giving green fodder to animals	4.30	4.11	4.28
5.	Preparing concentrate mixture	4.27	3.63	3.96
6.	Giving conc. mixture to animals	3.93	3.70	3.91
7.	Giving minerals/salts to animals	2.69	2.35	2.22
8.	Giving drinking water to animals	3.40	3.24	2.71
9.	Giving feed to animals	4.58	3.99	4.28
	Overall Mean Score	3.48	3.16	3.32

Calculated $'t'_{ab} = 5.54^{**}$, $'t'_{bc} = 0.37$, $'t'_{ca} = 3.56^{**}$
Tabulated $'t'_0$: $P < 0.05 = 2.31$; $P<0.01 = 3.36$

In order to find out if there was any significant difference between the role expectation, perception and performance in relation to the feeding activities, a paired 't' test was computed. Table 4.24 reveals that there was a significant difference (P<0.01) between the expectations of the spouse and the perception of the farm woman about her role in feeding the dairy animals, thereby, indicating that the farm women perceived their role in this segment differently than the expectations of the spouses. Table 4.24 further reveals that the performance of different feeding activities by the farm women was also significantly different (P<0.01) from the expectations of the spouse. There was no significant difference between the perception and performance of the different activities by the farm women, indicating that farm women perform their role in the feeding segment as per their own perceptions and were not influenced by the expectations of their spouses.

4.3.4 *Role Expectation : Breeding*

The data presented in Table 4.25 indicates that the expectations of the spouses from the farm women were low (overall mean score, 2.11) in the role segment of breeding. This indicated that spouses do not expect from the farm women to perform the activities related to breeding of the dairy cattle and buffaloes. The role segment breeding had been given the 4th rank by the judges (paired comparison technique) giving less weightage as compared to other role segments (Table 4.31).

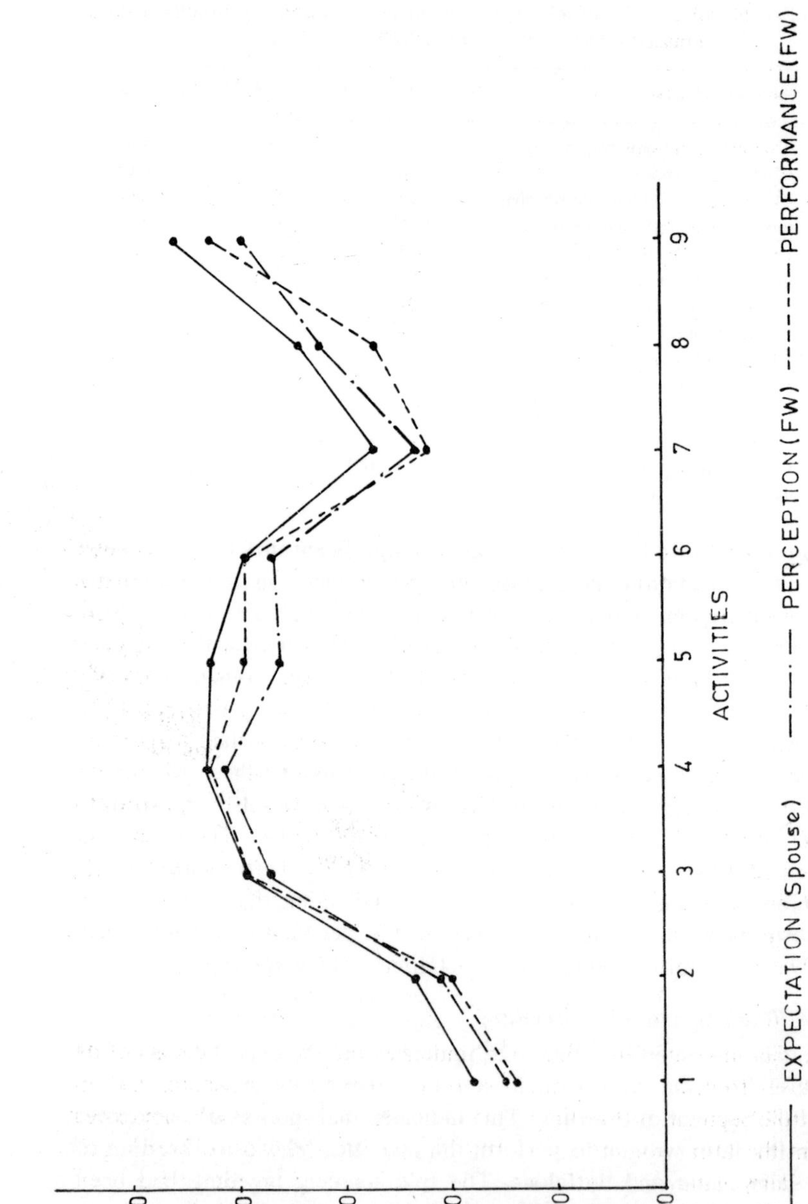

FIG 4.1: Expectation, perception and performance of feeding activities

Table 4.25 and Fig. 4.2 further indicate that only two activities viz., 'arranging materials at the time of parturition' and 'taking care of animals at the time of parturition', were highly expected by the spouses to be performed by the farm women.

4.3.5 *Role Perception : Breeding*

The perusal of Table 4.25 further reveals that perception of farm women in the role segment of breeding was low (overall mean score, 1.74). Table 4.25 also reveals that the activities of this role segment which were perceived to be required to be performed by her were, the same as they were expected by the spouses, such as 'taking care of animals at the time of parturition' and 'arranging materials at the time of parturition'.

TABLE 4.25: Mean scores of role expectation of spouses and role perception and performance of farm women – Breeding

Activities/Tasks	Expectation[s]	Perception[b]	Performance[c]
1. Observing animals for signs of heat	2.16	1.46	1.21
2. Taking the animals for natural service	1.12	1.06	1.05
3. Taking the animal for A.I.	1.08	1.03	1.03
4. Observing animal for symptomatic pregnancy after service/insemination	2.10	1.67	1.33
5. Taking animal for pregnancy diagnosis to Vety. hospital	1.09	1.28	1.08
6. Observing animal for reproductive problems during pregnancy	2.12	1.59	1.38
7. Taking the animal to Veterinarian for treatment of reproductive disorders	1.25	1.02	1.04
8. Taking care of animal at the time of parturition	3.72	3.25	2.97
9. Arranging materials at the time of parturition	4.75	3.92	3.88
10. Calling Vety. doctor in case of any problem in calving	1.32	1.09	1.07
11. Keeping watch on dropping of placenta after parturition	2.66	2.12	1.79
12. Disposal of the placenta	1.94	1.41	1.71
Overall Mean Score	2.11	1.74	1.61

Calculated $'t'_{ab} = 4.30^{**}$, $'t'_{bc} = 2.09$, $'t'_{ca} = 4.32$

Tabulated $'t'_{0}$: $P < 0.05 = 2.20$; $P < 0.01 = 2.72$

4.3.6 *Role Performance : Breeding*

The data in Table 4.25 points out that the performance of the breeding activities/tasks by farm women was very low (overall mean score, 1.61). The results were in consonance with the results reported by Dey (1977) that activities associated with breeding of dairy animals are performed by male members of the family.

The data in Table 4.25 and Fig. 4.2 clearly indicate that the farm women were not performing most of the activities in breeding segment of dairy animal production, except those which are related with the parturition of the animal. It was, however, interesting to note that she was not taking care of the animals at the time of parturition, which enjoyed high role expectations of spouses and high role perception of farm women up to the levels of their spouses' expectations or their own perceptions (mean score, 2.97). Her performance on 'arranging materials at the time of parturition was, however, high (mean score, 3.88) and in consonance with the expectations of the spouse and her own perception. All other activities were low on the expectations as well as perception and performance.

The paired 't' test (Table 4.25) revealed that there was highly significant difference ($P<0.01$) between expectations of the spouses and the perceptions of the farm women about her role in activities associated with breeding of the dairy animals, thereby, indicating that the farm women perceived their role in this segment differently than the expectations of the spouses. The role performance of farm women was also significantly different ($P<0.01$) from the expectations of the spouses. There was, however, no significant difference between the perception and performance of the different activities by the farm women, indicating that farm women performed their role in the breeding segment as per their own perceptions and their performance was not influenced by the expectations of their spouses.

4.3.7 *Role Expectation : Management*

Table 4.26 shows that the expectations of the spouses from the farm women in the role segment of management were high (overall mean score, 3.21). This indicated that spouses expected from the farm women to perform most of the activities associated with management of the dairy cattle and buffaloes. When the different activities/ tasks related to the management of dairy animals were studied, it was found

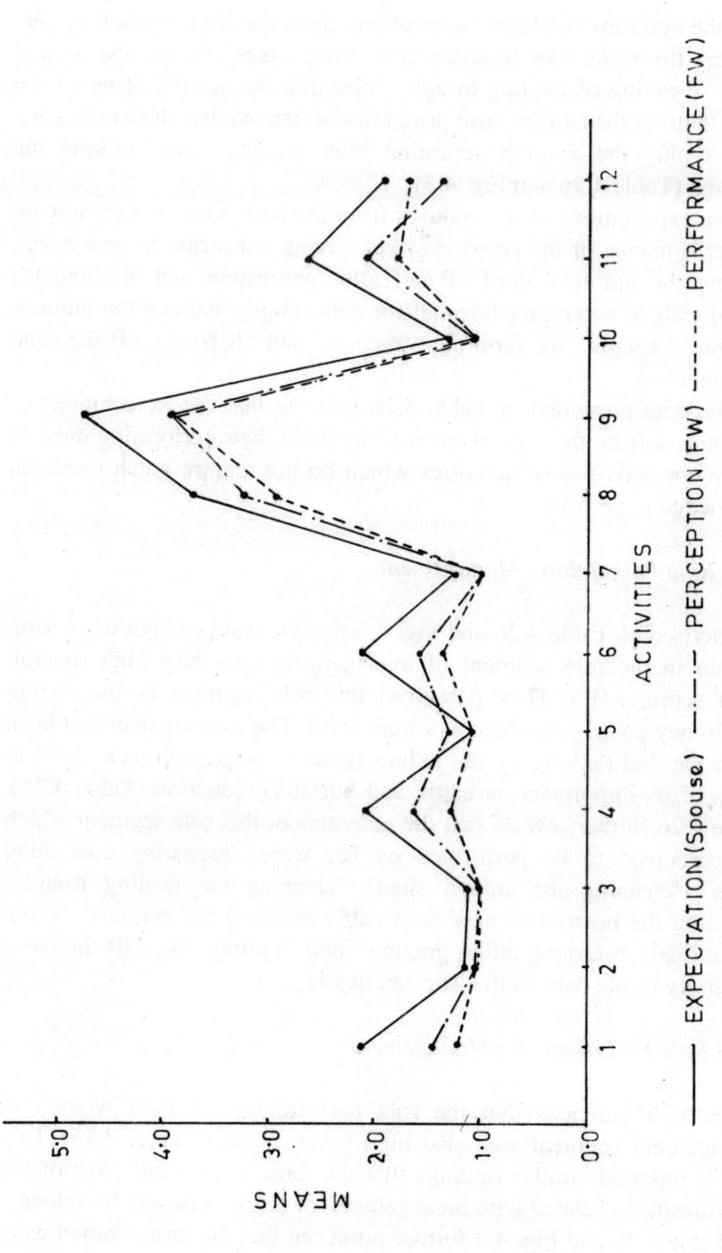

FIG 4.2: Expectation, perception and performance of breeding activities

that the spouses had high expectations from the farm women in performing the tasks like 'prepare cow dung cakes', 'clean the animal shed', 'cleaning of feeding trough', 'cleaning the nostrils of new born calf', 'putting the calf in close proximity of dam so that she can lick the calf', typing the animals returning after grazing' and 'milking the animals' (Table 4.26 and Fig. 4.3).

The expectations of the spouses from the farm women were low on the performance of the activities like, 'giving colostrum to new born', 'cutting the umblical cord of calf after parturition and dressing it', 'taking calf to veterinary hospital for dehorning', 'bathing the animals at home', 'keeping the farm/dairy records' and 'disposing off the dead animals'.

The data presented in Table 4.26 indicate that the expectations of the spouses from the farm women were about their performing most of the indoor activities or activities which do not require much technical knowledge to perform it.

4.3.8 *Role Perception : Management*

The perusal of Table 4.26 and Fig. 4.3 reveals that perceptions of farm women in the role segment of management were also high (overall mean score, 3.02). They perceived this role segment as the one in which they could contribute at a high level. The role segment had been given the 2nd ranking by the judges (paired comparison technique) in terms of its importance in cattle and buffalo production (Table 4.31). Table 4.26 further reveals that the activities of this role segment which she perceived to be performed by her were, 'preparing cow dung cakes', 'cleaning the animal shed', 'cleaning the feeding trough', 'cleaning the nostrils of new born calf', 'milking the animals', 'tying the animals returning after grazing' and 'putting the calf in close proximity of the dam so that she can lick it'.

4.3.9 *Role Performance : Management*

Table 4.26 indicates that the role performance of farm women in management segment was also high (overall mean score, 3.15). Dey (1977) reported similar findings that the farm women did most of the activities/tasks related with management of dairy cattle and buffaloes.

Table 4.26 and Fig. 4.3 further point out that the farm women were performing at a high level the activities such as 'preparing cow dung cakes', 'cleaning the animal shed', 'milking the animals', 'typing the animals returning after grazing', 'cleaning the feeding trough' and 'putting the calf in close proximity of the dam so that she can lick it'.

Table 4.26 further reveals that no significant difference ($P<0.05$) was found between role expectation of spouses, role perception and performance of farm women indicating that farm women were performing the activities in management of dairy cattle and buffaloes as their perception and expectations of the spouses.

TABLE 4.26: Mean scores of role expectation of spouses and role perception and performance of farm women — Management

Activities/Tasks	Expectation[a]	Perception[b]	Performance[c]
1. Giving colostrum to newborn calf	2.51	1.39	1.08
2. Cutting the umblical cord of calf after parturition and dressing it	2.73	2.59	2.75
3. Cleaning the nostrils and body of newborn calf	4.42	3.98	4.98
4. Putting the calf in close proximity of dam so that she can lick the calf	3.37	3.26	3.85
5. Taking calf to Vety. hospital for dehorning	1.25	1.38	1.01
6. Cleaning the animal shed	4.72	4.47	4.85
7. Preparing cow dung cakes	4.94	4.91	4.95
8. Cleaning the feeding trough/ manger	4.55	3.99	4.58
9. Tie the animals returning after grazing	3.77	3.85	3.90
10. Bathing the animals at home	2.66	2.69	2.68
11. Keeping the farm/dairy records	1.65	1.77	1.01
12. Milking the animals	4.06	3.68	4.25
13. Disposing off the dead animals	1.14	1.29	1.02
Overall Mean Score	3.21	3.02	3.15

Calculated $'t'_{ab} = 1.93$, $'t'_{bc} = 0.94$, $'t'_{ca} = 0.47$
Tabulated $'t'_{o}$: $P < 0.05 = 2.18$, $P < 0.01 = 3.06$

4.3.10 *Role Expectation : Health Care*

Perusal of Table 4.27 points out that the expectations of the spouses from the farm women were low (overall mean score, 2.55) in the role segment of health care. This indicated that spouses do not expect from the farm women to perform most of the activities associated with the health care of the dairy cattle and buffaloes.

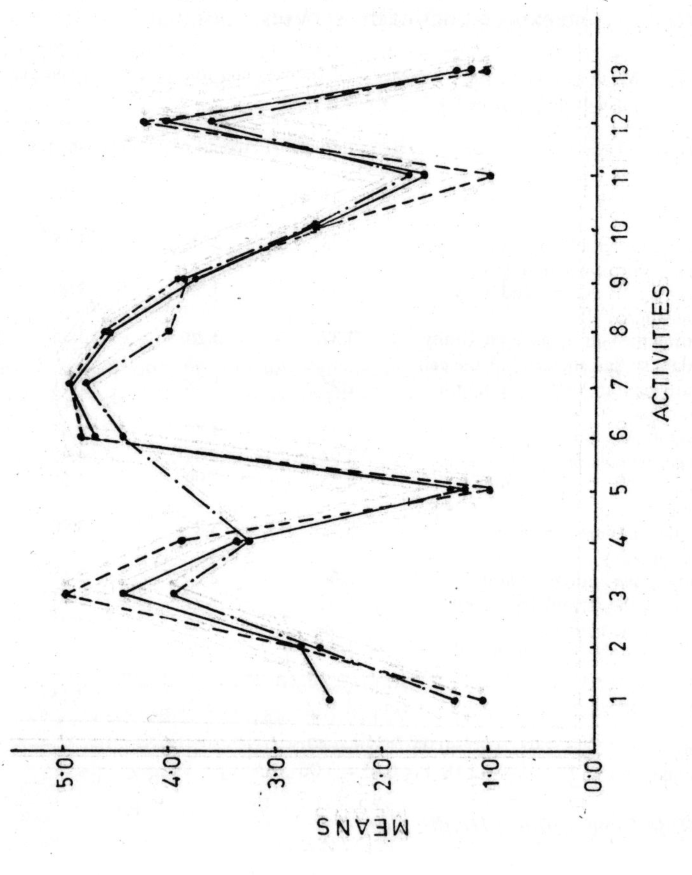

FIG 4.3: Expectation, perception and performance of management activities

Table 4.27 and Fig. 4.4 further reveal that only three activities viz., 'providing first aid in case of emergency', 'arranging household medications for sick animals' and 'taking care of sick animals', were expected by the spouses to be performed by the farm women in this segment.

The spouses did not expect the farm women to perform at high level the activities like 'observing animal for any symptom of sickness', 'taking sick animals to Vety. hospital for treatment', 'isolating the sick animals, if necessary', 'administering/applying medicines to the sick animals', 'getting the animal vaccinated against contagious diseases', 'observing ectoparasites in animals at intervals', 'calling very. doctor in case of emergency' and 'getting the animal dewormed regularly'.

4.3.11 *Role Perception : Health Care*

Table 4.27 and Fig. 4.4 points out that perception of farm women in role segment of health care was also low (overall mean score, 2.13). The role segment had been given 3rd ranking by the judges (paired comparison technique) in terms of its importance in cattle and buffalo production (Table 4.31). Table 4.27 and Fig. 4.4 further reveal that the activities which she perceived to be performed by her were, 'providing first aid in case of emergency', 'arranging household medications for sick animals' and 'taking care of sick animals'. It will be observed that these activities are indoor activities requiring relatively less physical labour and energy.

4.3.12 *Role Performance : Health Care*

Table 4.27 reveals that performance of the farm women was found to be low (overall mean score, 1.95) in the segment of health care of dairy cattle and buffaloes. Dey (1977) also reported similar results that males did majority of the activities/tasks associated with health care.

The data in Table 4.27 and Fig. 4.4 further reveal that farm women were not performing most of the activities/tasks in this segment except in two activities where they were contributing to the health care of the dairy animals. They were, 'arranging household medications for sick animals' and 'providing first-aid in case of emergency'. But the role performance was found low in case of 'taking care of sick animals' which was having high role expectation of spouses as well as high role perception of farm women.

TABLE 4.27: Mean scores of role expectation of spouses and role perception and performance of farm women — Health care .

Activities/Tasks	Expectation[a]	Perception[b]	Performance[c]
1. Observing animal for any symptom of sickness	2.85	2.54	2.32
2. Taking sick animal to Vety. hospital for treatment.	1.38	1.26	1.16
3. Isolating the sick animals, if necessary	2.25	1.72	1.65
4. Providing first aid in case of emergency	4.01	3.31	3.12
5. Arranging household medication for sick animals	4.76	4.08	3.93
6. Administering/applying medicines to the sick animals	2.41	1.73	1.59
7. Getting the animal vaccinated against contagious diseases	1.87	1.12	1.09
8. Taking care of sick animals	3.55	3.06	2.95
9. Observing the ectoparasites in animal at intervals	2.32	2.32	1.37
10. Calling Vety. doctor in case of emergency	1.16	1.02	1.04
11. Getting the animal dewormed regularly	1.47	1.33	1.05
Overall Mean Score	2.55	2.13	1.95

Calculated $'t'_{ab} = 4.92^{**}$, $'t'_{bc} = 2.55^{*}$, $'t'_{ca} = 7.44^{**}$
Tabulated $'t'_{o}$: $P < 0.05 = 2.23$, $P < 0.01 = 3.17$

Paired 't' test (Table 4.27) indicated that significant difference (P<0.01) was found between role expectation of spouses and role perception of farm women. The difference was also found significant (P<0.01) between role expectation of spouses and role performance of farm women.

Table 4.27 further reveals that there was also significant difference (P<0.05) in the paired 't' values between role perception and role performance of farm women in the tasks associated with health care indicating that the role expectation of spouses, role perception and performance of farm women significantly differed from each other. What was expected by the spouses was neither perceived nor performed by farm women. Further, the farm women did not even perform the activities which they perceived should be performed by them.

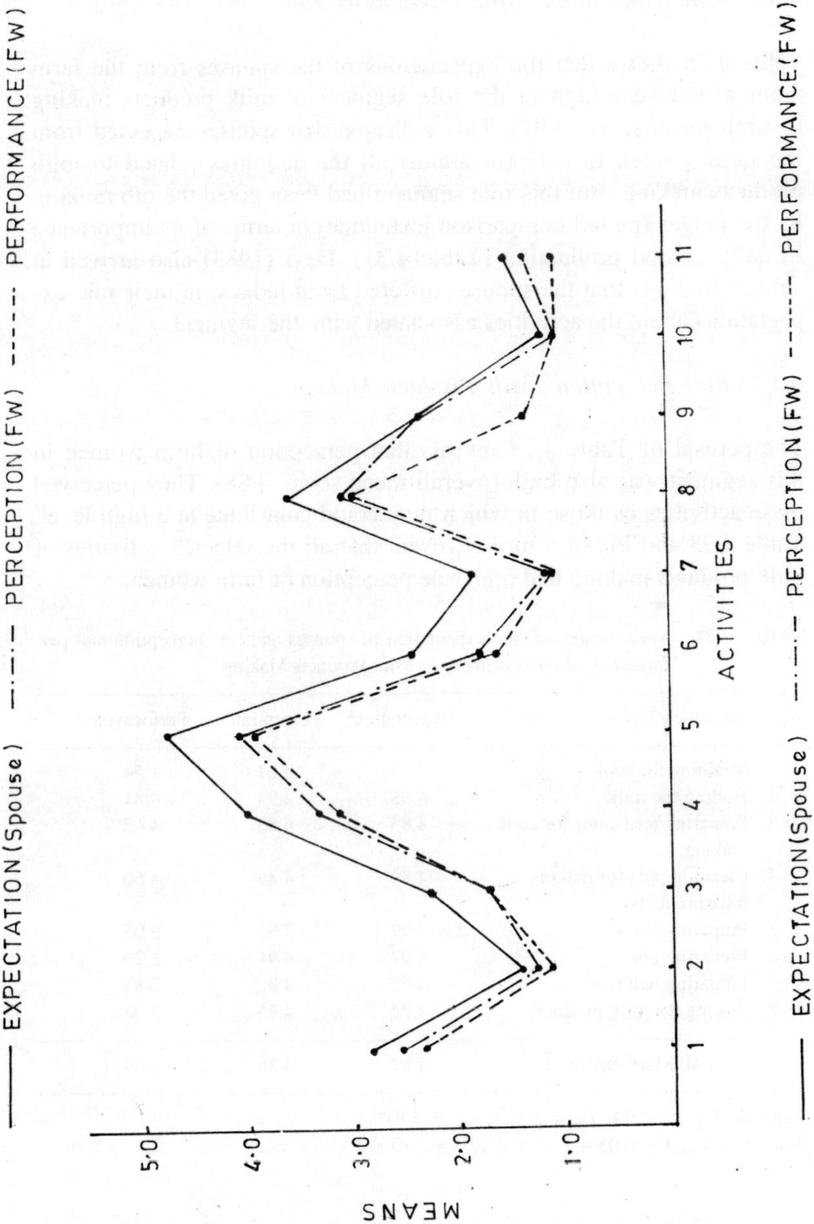

FIG 4.4: Expectation, perception and performance of health care activities

4.3.13 *Role Expectation : Milk Products Making*

Table 4.28 shows that the expectations of the spouses from the farm women was very high in the role segment of milk products making (overall mean score, 4.92). This indicated that spouses expected from the farm women to perform almost all the activities related to milk products making. But this role segment had been given the 6th ranking by the judges (paired comparison technique) in terms of its importance in dairy animal production (Table 4.31). Devi (1983) also arrived at similar findings that the spouses differed from judges, in their role expectations about the activities associated with the segment.

4.3.14 *Role Perception : Milk Products Making*

The perusal of Table 4.28 reveals that perception of farm women in this segment was also high (overall mean score, 4.88). They perceived these activities as those in which they could contribute at a high level. Table 4.28 and Fig. 4.5 further reveal that all the selected activities of milk products making had high role perception of farm women.

TABLE 4.28: Mean scores of role expectation of spouses and role perception and performance of farm women — Milk Products Making

Activities/Tasks	Expectation[a]	Perception[b]	Performance[c]
1. Straining the milk	4.87	4.67	1.54
2. Boiling the milk	4.95	4.93	4.81
3. Preparing fermenter for curd making	4.85	4.83	4.63
4. Cleaning pots for making milk products	4.89	4.89	4.60
5. Preparing *khoa*	4.99	4.91	3.05
6. Preparing *ghee*	4.97	4.91	3.20
7. Preparing *makhan*	4.92	4.92	2.83
8. Storing the milk products	4.94	4.95	4.08
Over all Mean Score	4.92	4.88	3.59

Calculated $'t'_{ab} = 2.04$, $'t'_{bc} = 3.34^*$, $'t'_{ca} = 3.30^*$
Tabulated $'t'_{o}$: $P < 0.05 = 2.37$, $P < 0.01 = 3.50$

4.3.15 *Role Performance : Milk Products Making*

The data on role performance in Table 4.28 shows that the farm women were performing, at a high level, most of the activities related to milk products making (overall mean score, 3.59), except two activities i.e. 'straining the milk' and 'preparing makhan (Desi butter)' in which their performance was very low and low, respectively, in comparison to the expectations.

Table 4.28 emphasised that significant differences ($P < 0.05$) were found between role perception and role performance of farm women and between role expectation of spouses and role performance of farm women, indicating that farm women were performing significantly differently as compared to their own perceptions and the expectations of spouses in this segment.

No significant differences ($P > 0.05$) were, however, found between the expectation of spouses and perception of farm women in the role segment, indicating that farm women perceived their roles in milk products making as their spouses expected from them.

4.3.16 *Role Expectations : Green Fodder Production*

The data in Table 4.29 emphasises that the role expectation of spouses from farm women was low (overall mean score, 2.05) in the role segment of green fodder production. This indicates that spouses do not expect from the farm women to perform the activities associated with green fodder production. This role segment had been given 5th ranking by the judges (paired comparison technique) indicating less weightage as compared to other role segments (Table 4.31). The activity/task wise means indicate that the spouses expected the farm women to be associated with only four activities viz., 'helping in preparing field for fodder', 'helping in hoeing and weeding the fodder crop', 'helping in sowing the fodder crop' and 'helping in cutting the fodder crop' (Table 4.29 and Fig. 4.6).

4.3.17 *Role Perception : Green Fodder Production*

A perusal of Table 4.29 points out that perception of farm women in the role segment of green fodder production was also low (overall mean score, 1.84). Table 4.29 and Fig. 4.6 further reveal that the activities which were perceived to be performed at a higher level by the farm women were, the same as were expected by the spouses, except 'helping in hoeing and weeding the fodder crop' which was low in perception.

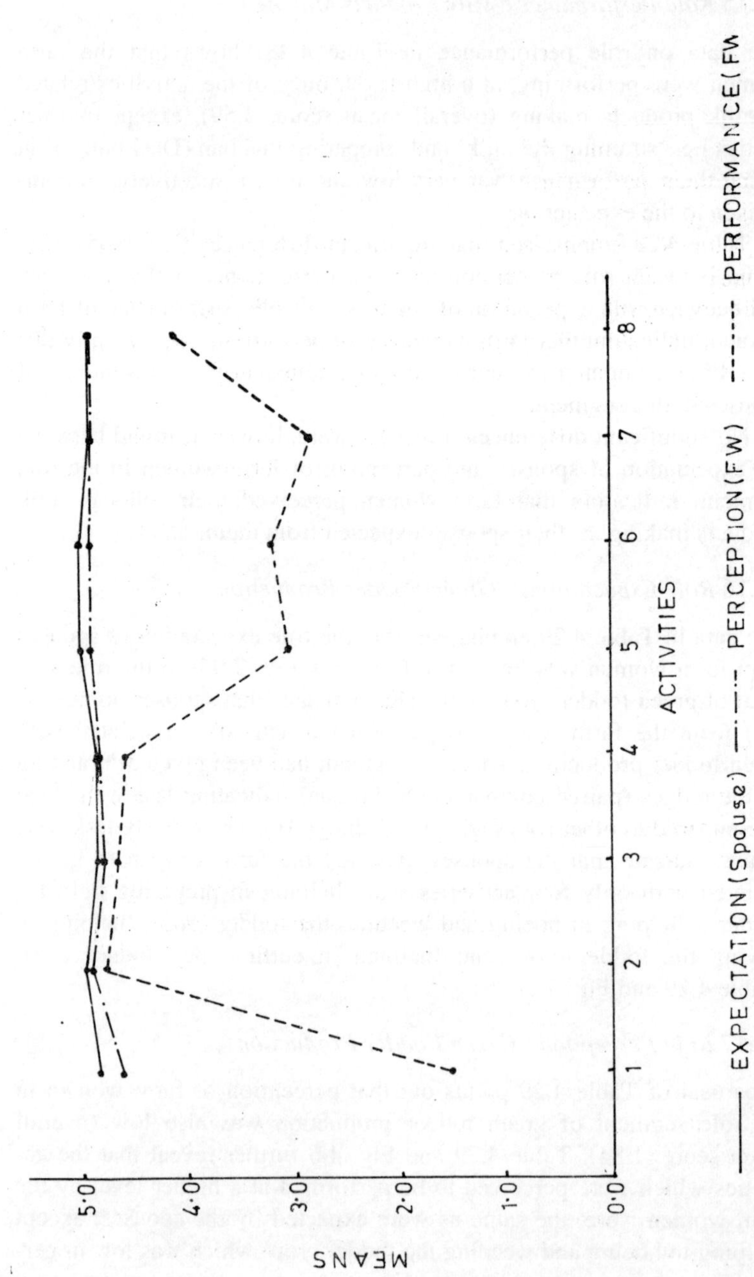

FIG 4.5: Expectation. perception and performance of milk products making activities

4.3.18 *Role Performance : Green Fodder Production*

The data in Table 4.29 indicate that role performance by farm women was also low (overall mean score, 1.98) in this segment.

The data in Table 4.29 and Fig. 4.6 clearly indicate that the farm women were performing most of the activities in green fodder production at a low level except, 'helping in preparing field for fodder', 'helping in sowing the fodder crop', 'helping in hoeing and weeding the fodder crop' and 'helping in cutting the fodder crop'. It is obvious that they were performing these activities at the level expected by their spouses.

Table 4.29 emphasises that significant difference ($P<0.05$) was found between role expectation of spouses and role perception of farm women in the role segment of green fodder production indicating that farm women perceived their role in this segment differently than the expectations spouses.

TABLE 4.29: Mean scores of role expectation of spouses and role perception and performance of farm women — Green Fodder Production

	Activities/Tasks	Expectation[a]	Perception[b]	Performance[c]
1.	Preparing field for fodder	1.72	1.28	1.25
2.	Helping in preparing field for fodder	3.41	3.24	3.02
3.	Arranging fodder seeds	1.55	1.19	1.03
4.	Sowing fodder crop	1.41	1.19	1.10
5.	Helping in sowing the fodder crop	3.08	3.01	3.15
6.	Hoeing and weeding the fodder crop	2.20	1.90	1.95
7.	Helping in hoeing and weeding the fodder crop	3.21	2.43	3.11
8.	Irrigating the fodder crop	1.26	1.22	1.25
9.	Transporting the green fodder from field	1.53	1.72	2.21
10.	Cutting the green fodder	1.74	1.88	2.09
11.	Helping in cutting the fodder crop	3.19	3.19	3.07
12.	Cutting the grasses/tree leaves from forest for fodder	1.72	1.56	2.56
13.	Making the silage	1.33	1.10	1.04
14.	Making the hay	1.33	1.09	1.02
	Overall Mean Score	2.05	1.84	1.98

Calculated $'t'_{ab} = 2.92^*$, $'t'_{bc} = 1.72$, $'t'_{ca} = 0.53$
Tabulated $'t'_{o}$: $P < 0.05 = 2.16$, $P < 0.01 = 3.01$

Table 4.29 further reveals that no significant difference was found between role perception and role performance of farm women and between role expectation of spouses and role performance of farm women indicating that farm women were performing the activities in green fodder production in accordance with their role perception and role expectation of spouses.

4.3.19 *Role Expectation : Marketing*

Perusal of Table 4.30 points out that the expectation of the spouses from the farm women was very low (overall mean score, 1.53) in role segment of marketing. This indicated that spouses do not expect from the farm women to perform the activities associated with marketing. This role segment had been given 7th ranking by the judges (paired comparison technique) giving least importance to it in dairy animal production as compared to other segments (Table 4.31).

Table 4.30 and Fig. 4.7 clearly points out that the farm women were not expected by their spouses to perform most of the activities in role segments marketing, except the 'sale of milk to the milkman', who usually comes to their own door.

4.3.20 *Role Perception : Marketing*

Table 4.30 shows that perception of farm women in the role segment of marketing was also low (overall mean score, 1.45). Table 4.30 and Fig. 4.7 further reveal that the activities of this role segment which were perceived to be required to be performed by her were, 'selling the milk to milkman in village' and 'selling milk products in the village', although the expectation of the spouse on the later activity was low.

4.3.21 *Role Performance : Marketing*

The data in Table 4.30 indicate that role performance for practically all the activities was very low (overall mean score, 1.19) in this role segment.

Perusal of Table 4.30 and Fig. 4.7 reveals that all the activities in role segment of marketing were performed at a very low level by the farm women. Even the activities which were expected by their spouses and the activities which were perceived by farm women to be performed by them, were also performed by them at a very low level.

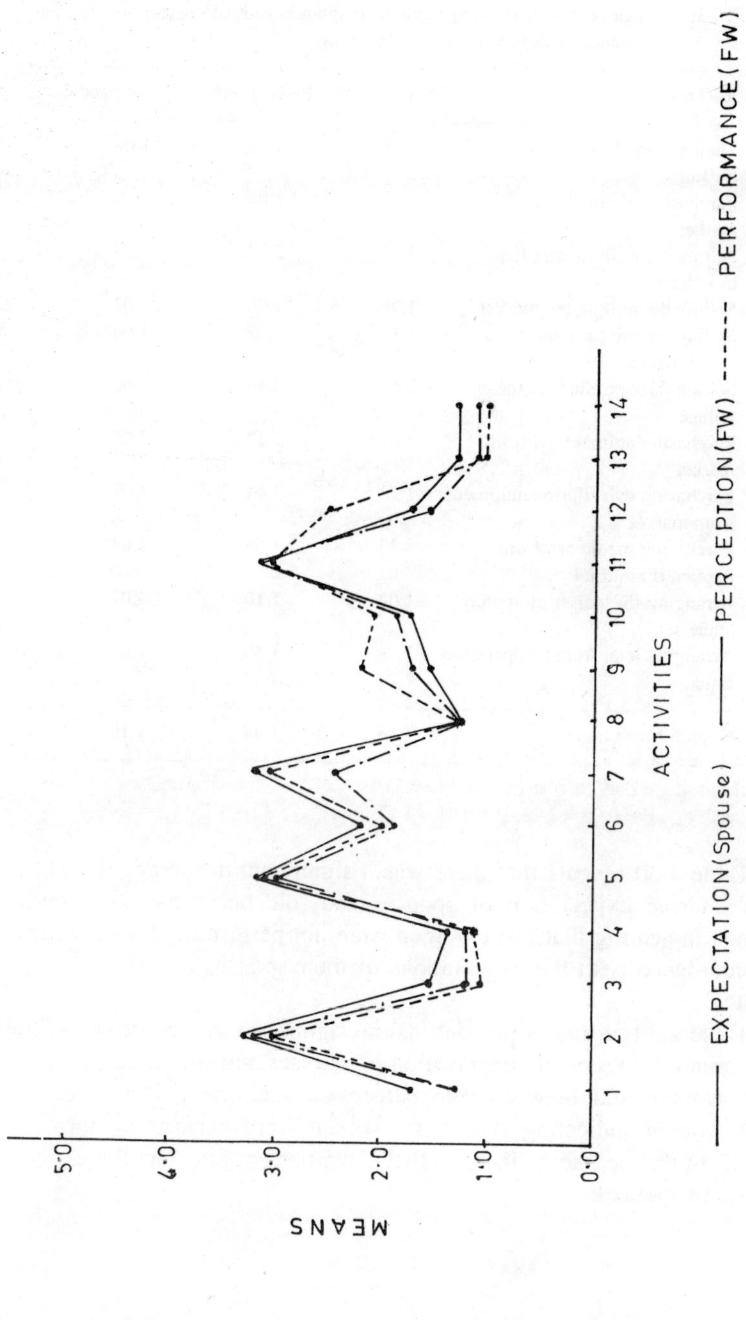

FIG 4.6: Expectation, perception and performance of green fodder production activities

TABLE 4.30: Mean scores of role expectation of spouses and role perception and performance of farm women — Marketing

Activities/Tasks	Expectation[a]	Perception[b]	Performance[c]
1. Taking animals to market for sale	1.28	1.06	1.08
2. Purchasing animals from market	1.19	1.04	1.03
3. Selling the milk to milk man in village	3.01	3.23	2.77
4. Selling the milk in the market	1.96	1.08	1.02
5. Selling the milk products in the market	1.18	1.02	1.04
6. Selling milk products in the village	2.52	3.19	1.06
7. Purchasing animal feed from market	1.21	1.12	1.05
8. Purchasing dairy/farm equipments from market	1.13	1.04	1.01
9. Purchasing medicines from market, if required	1.22	1.06	1.03
10. Arranging for loan from money lenders	1.08	1.10	1.01
11. Arranging loan from Cooperative Banks	1.08	1.30	1.01
Overall Mean Score	1.53	1.45	1.19

Calculated $'t'_{ab} = 0.49$, $'t'_{bc} = 1.50$, $'t'_{ca} = 2.55^*$
Tabulated $'t'_o : P < 0.05 = 2.23, P < 0.01 = 3.17$

Table 4.30 reveals that there was a significant difference (P<0.05) between role expectation of spouses and role performance of farm women indicating that farm women were not performing the activities in accordance with the expectations of their spouses in this role segment.

Table 4.30 further points out that no significant difference (P>0.05) was found between role expectation of spouses and role perception of farm women and between role perception and role performance of farm women indicating that farm women were performing the activities in this segment closer to their own perceptions than the expectations of spouses.

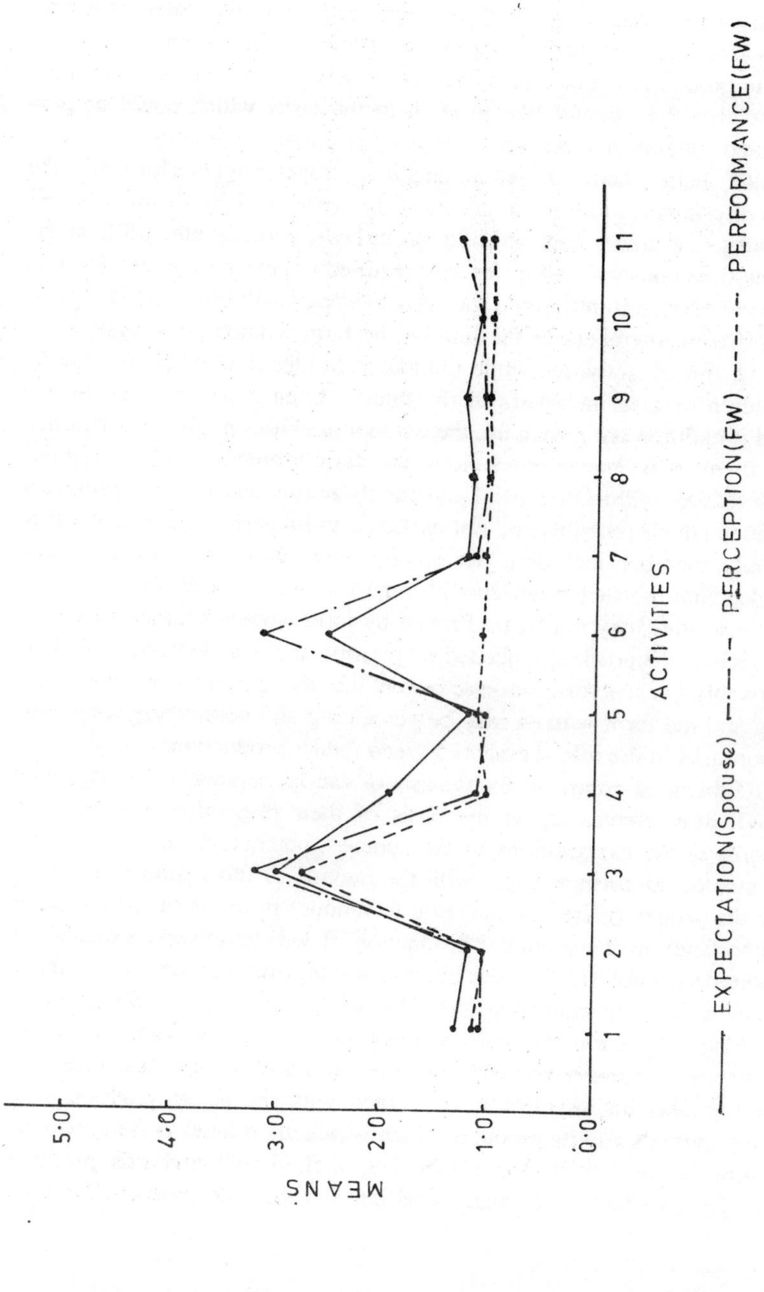

FIG 4.7: Expectation, perception and performance of marketing activities

It can be concluded from above results that the performance of the different activities/tasks were broadly in consonance with how the farm women themselves had perceived and interpreted their contributions for different activities/tasks of different role segment of dairy cattle production. Their performance of different activities/ tasks indicated that they would like to stick to the tasks which could be performed indoors and require less physical labour, probably because it would help them in simultaneously supervising/performing the activities/responsibilities expected to be performed by them as home-makers like taking care of children, in-laws, cooking etc. while at the same time conserving their energy required to perform these tasks. It is obvious that in joint family, the activities/responsibilities are delineated to different members of the family, the farm women performing most of the indoor activities while outdoor activities may be performed by children or male members of the family. Even if it is nuclear family and if children are grown up, the outdoor activities might be performed by them. It is, however, obvious that farm women, as home-makers, like to stay within the social and family norms and perform those activities which probably are not expected to be performed by any other family member including her spouse. Her responsibilities as home-maker may not leave her enough time for outdoor activities. Some of the activities may not be performed by farm women because they lack in technical knowledge needed to perform the activity/task well. It is probably because of economic reason that the spouses may be expecting and the farm women may be perceiving and performing some outdoor tasks in the role segment of green fodder production.

Table 4.31 compares the ranking of various segments of dairy cattle production arrived at on the basis of their respective overall mean scores of the expectations of the spouses, perceptions and the performance of the farm women, with the ranking of the segments assigned by the judges (paired comparison technique) in terms of their relative importance in dairy animal production. It will be observed that while judges assigned feeding the highest ranking the spouses expected the farm women do contribute the most in the segment of milk products making. The farm women's perceptions and the performance were, however, in consonance with the expectations of the spouses. This may be because this segment is concerned with the actual production of dairy animals and its products which is directly related to the economic returns to the family through the sale of fluid milk and milk products like *Ghee* and *Khoa*. In many rural families the milk products like curd constitute an essential part of the family's daily diet. The experts had ranked this segment at the 6th place. Because of this shifting of ranking all other rankings by the spouses and farm women above this segment moved down by one step maintaining the same sequential order which was assigned by the judges to these segments. Marketing was ranked the lowest by the judges as well as by the spouses and farm women.

TABLE 4.31: Comparative rankings of role segments by experts, in terms of importance, and by spouses and farm women, in terms of expectations, perceptions and performance

Role Segment	Paired comparison technique		Overall Mean Scores					
	Scale value	Rank	Expectation		Perception		Performance	
			OMS	Rank	OMS	Rank	OMS	Rank
Feeding	1.441	I	3.48	II	3.16	II	3.32	II
Management	1.252	II	3.21	III	3.02	III	3.15	III
Health care	1.037	III	2.55	IV	2.13	IV	1.95	V
Breeding	0.737	IV	2.11	V	1.74	V	1.61	VI
Green Fodder Production	0.713	V	2.05	VI	1.84	VI	1.98	IV
Milk Products	0.163	VI	4.92	I	4.88	I	3.59	I
Marketing	0.000	VII	1.53	VII	1.45	VII	1.19	VII

OMS = Overall Mean Score.

4.4 Differences Between Role Expectation, Perception and Performance According to Different Categories of Socio-economic Status

This part deals with the differences in role expectations of spouses, role perception and role performance of farm women according to the socio-economic status of the respondents.

To find out the significant difference, if any, between the role expectation of spouses, role perception and role performance of farm women, average mean scores for each activity in each socio-economic category were used for the analysis of variance. Critical difference was computed for those role segments and aspect of the role expectation, perception and performance in which value of 'F' was found significant in the analysis of variance.

4.4.1 *Role Expectation : Feeding*

The analysis of variance (Table 4.32) revealed that the computed 'F' value was highly significant (P<0.01) indicating that the role expectations of spouses of different socio-economic categories were statistically highly different. The spouses from different categories of socio-economic status expected differently the role farm women should play in feeding the dairy animals.

TABLE 4.32: Analysis of variance in role expectation of spouses, role perception and performance of farm women in feeding segment among different socio-economic categories

Source of variation	d.f.	Mean Squares		
		Expectation	Perception	Performance
Between groups	4	119.226 (5.832**)	29.050 (1.766)	42.3888 (3.643**)
Within groups (Error)	115	20.440	17.325	17.209

Note: Calculated 'F' values shown in parentheses.
** Significant at 1 per cent level of significance.

4.4.2 *Role Perception : Feeding*

The data in Table 4.32 further show that the computed value of 'F' for role perception by farm women of this segment was not significant. It could, therefore, be concluded that there was no difference in role perception of farm women from different socio-economic categories with regard to feeding the dairy cattle and buffaloes.

4.4.3 *Role Performance : Feeding*

The data in Table 4.32 further indicate that the calculated value of 'F' for role performance of farm women among different categories of socio-economic status, was highly significant (P<0.01). Hence, it could be concluded that there was a significant difference in role performance of farm women from different categories of socio-economic status with regard to feeding the dairy animals.

The critical difference was computed in order to find out the differences in role expectation of the spouses and role performance of farm women belonging to different socio-economic strata in relation to the segment of feeding the dairy animals as a significant difference had been found among the different categories in the analysis of variance.

TABLE 4.33: Critical difference in role expectation of spouses and performance of farm women in feeding segment among different socio-economic categories

	Categories of socio-economic status				
	Lower	Lower-middle	Middle	Upper-middle	Upper
Role expectation					
Average	32.50[a]	31.52[a]	31.00[a]	26.00[b]	24.60[b]
S.E.	±3.19	±1.36	±0.69	±0.58	±2.02
Role perception					
Average	—	—	—	—	—
S.E.	—	—	—	—	—
Role performance					
Average	33.00[a]	30.26[a]	30.52[a]	26.55[b]	29.34[ab]
S.E.	±2.93	±0.53	±0.64	±1.25	±1.86

Note: The averages bearing different superscripts are significantly different at 5 per cent level of significance.

The pairwise comparison of average role expectation of spouses belonging to different socio-economic categories (Table 4.33) indicated that lower, lower-middle and middle categories of socio-economic status did not differ significantly ($P<0.05$) from each other. Same was true between upper-middle and upper categories of socio-economic strata, which indicated that role expectations of spouses in activities associated with feeding of the dairy cattle and buffaloes was of the same order among the spouses of those two clusters of socio-economic strata.

The average role expectation of spouses in the feeding segment was, however, significantly higher ($P<0.05$) in lower, lower-middle and middle categories than in middle and upper categories of socio-economic status, which included that spouses from upper-middle and upper categories of socio-economic status differ in their role expectation in this segment than the spouses from lower, lower-middle and middle categories.

The data in Table 4.33 reveals that in the families belonging to the lower socio-economic status the spouses expected a higher participation of farm women in the activities related with the feeding of dairy animals. In families belonging to the higher socio-economic status the expectations on the participation of their women folk in the dairy animal feeding activities was lower than in lower socio-economic strata. It is, therefore, obvious that higher the socio-economic status of the family lower is the role expectations from the farm women in dairy animal feeding activities. Apart form the economic viability of the family, the social values and norms appear to have some bearing on this difference in the expectations about the participation of farm women in feeding activities.

Table 4.33 further points out that the lowest performance of farm women in the role of feeding the dairy cattle and buffaloes was observed in upper-middle category and this value was significantly ($P<0.05$) lower than the average role performance values in lower, lower-middle and middle categories of socio-economic status. The values in upper-middle category did not show a significant difference ($P>0.05$) from upper category of socio-economic status. Lower, lower-middle, middle and upper categories also did not show a significant difference among them and were thus of the same order.

It can be concluded that farm women from lower socio-economic status category performed their role in feeding the dairy animals at a higher level than those belonging to the higher socio-economic status. This may be because in lower socio-economic status category the the spouses are most of the time away from home, working in the fields or earning a livelihood and farm women have to take care of most of the tasks at home.

4.4.4 *Role Expectation : Breeding*

Perusal of Table 4.34 indicates that the computed 'F' value for role expectation of spouses about the role of the farm women in activities associated with breeding of dairy animals among different categories of socio-economic status was highly significant ($P<0.01$) indicating that there was a significant difference in role expectation of spouses about the farm women's participation in the tasks related with the breeding of the dairy animals.

TABLE 4.34: Analysis of variance in role expectation of spouses, role perception and performance of farm women in breeding segment among different socio-economic categories

Source of variation	d.f.	Mean Squares		
		Expectation	Perception	Performance
Between groups	4	57.960 (3.962**)	29.940 (2.967*)	23.829 (1.996)
Within groups (Error)	115	17.854	11.515	11.936

Note: Calculated 'F' values shown in parentheses.
** Significant at 1 per cent of significance.
* Significant at 5 per cent of significance.

4.4.5 *Role Perception : Breeding*

Perusal of Table 4.34 points out that computed 'F' value for role perception of farm women in breeding tasks of the dairy animals, was statistically significant ($P<0.05$). It could, therefore, be concluded that in activities associated with breeding of dairy animals there was a significant different in role perception of farm women from different socio-economic strata.

4.4.6 *Role Performance : Breeding*

Table 4.34 further shows that the computed 'F' value for the performance of activities in breeding segment was not significant. Hence, it could be concluded that there was no difference in role performance of farm women from different socio-economic categories in breeding activities of the dairy cattle and buffaloes.

To find out the difference in the average values in role expectation of the spouses, role perception and role performance of farm women belonging to different socio-economic strata, the critical differences were calculated.

Critical difference analysis (Table 4.35) revealed that role expectation of spouses about farm women in activities associated with breeding of dairy animals was significantly high ($P<0.05$) in upper-middle than lower-middle and middle socio-economic categories but was not significantly ($P>0.05$) different from lower and upper categories. The values of lower, upper-middle and upper categories of socio-economic status, also did not differ statistically ($P>0.05$) among themselves.

TABLE 4.35: Critical difference in role expectation of spouses and role perception of farm women in breeding segment among different socio-economic categories

	Categories of socio-economic status				
	Lower	Lower-middle	Middle	Upper-middle	Upper
Role expectation					
Average	26.60^{ab}	23.98^{b}	25.45^{b}	28.45^{a}	28.00^{ab}
S.E.	±2.99	±1.27	±0.65	±0.54	±1.89
Role perception					
Average	20.58^{ab}	19.24^{b}	21.50^{bc}	23.79^{ac}	25.55^{c}
S.E.	±2.39	±0.52	±1.52	±0.43	±1.02
Role performance					
Average	—	—	—	—	—
S.E.	—	—	—	—	—

Note: The averages bearing different superscripts are significantly different at 5 per cent level of significance.

It can be concluded that role expectations of spouses from upper-middle and upper socio-economic categories about farm women in breeding activities of dairy animals was high, since the spouses in higher socio-economic status families due to change in social values and realisation of need, expect from their wives to perform higher role in this segment.

Table 4.35 further reveals that the farm women belonging to the upper category of socio-economic status perceived their role in this segment differently ($P<0.05$) than those belonging to the lower and lower-middle socio-economic strata. The perceptions of the farm women from the upper-middle strata were also statistically significantly different ($P<0.05$) from the farm women belonging to lower-middle strata.

It can be concluded that farm women in higher socio-economic strata perceived their role in breeding activities at a higher level than those in the lower socio-economic status categories.

4.4.7 *Role Expectation: Management*

Table 4.36 indicates that the expectations of spouses of different socio-economic categories from farm women (F=5.734) in activities associated with management of dairy animals were highly significantly different (P<0.01), indicating that a significant difference exists among different categories in role expectation from farm women in managemental activities of dairy animals.

TABLE 4.36: Analysis of variance in role expectation of spouses, role perception and performance of farm women in management segment among different socio-economic categories

Source of variation	d.f.	Mean Squares		
		Expectation	Perception	Performance
Between groups	4	86.101 (5.734**)	20.230 (1.696)	10.167 (1.132)
Within groups (Error)	115	16.100	11.923	8.974

Note: Calculated 'F values shown in parentheses.
** Significant at 1 per cent level of significance.

4.4.8 *Role Perception: Management*

Table 4.36 shows that the computed 'F' value for role perception of farm women was not significant. Therefore, it could be concluded that there was no difference in the perception of farm women from different socio-economic categories about their role in the activities associated with the management of dairy animals.

4.4.9 *Role Performance : Management*

The data in Table 4.36 further shows that the computed 'F' value for role performance of farm women in the activities related to management of dairy animals by the farm women belonging to different categories of socio-economic status, was not significant. It could, therefore, be concluded that there was no difference in the performance of farm women from different socio-economic strata in this role segment.

The critical difference was computed in order to find out the difference in averages in the role expectation of spouses, role perception and performance of farm women among different categories of socio-economic status in management of dairy animals (Table 4.37).

TABLE 4.37: Critical difference in role expectation of spouses in management segment among different socio-economic categories

	Categories of socio-economic status				
	Lower	Lower-middle	Middle	Upper-middle	Upper
Role expectation					
Average	44.39^{ab}	44.91^{a}	43.50^{ab}	41.09^{b}	39.66^{b}
S.E.	±2.83	±0.52	±0.62	±1.20	±1.78
Role perception					
Average	—	—	—	—	—
S.E.	—	—	—	—	—
Role performance					
Average	—	—	—	—	—
S.E.	—	—	—	—	—

Note: The averages bearing different superscripts are significantly different at 5 per cent level of significance.

The role expectation of spouses about farm women in management of dairy cattle and buffaloes, was found significantly higher ($P<0.05$) in lower-middle category than upper-middle and upper socio-economic categories (Table 4.37). The group differences in management of dairy animals revealed that role expectation of spouses from farm women was more in lower socio-economic category than in higher socio-economic strata. No significant difference in the expectations was found between other socio-economic categories.

4.4.10 *Role Expectation : Health Care*

The computed 'F' value was highly significant ($P<0.01$) for the expectations of spouses from the farm women in health care activities of dairy cattle and buffaloes among the different socio-economic categories (Table 4.38).

4.4.11 *Role Perception : Health care*

The data presented in Table 4.38 pointed out that the computed 'F' value for role perception of farm women for the health care activities of dairy cattle and buffaloes was highly significant ($P<0.01$). It could, therefore, be concluded that there was a significant difference in role perception of farm women from different socio-economic categories about the health care activities of dairy animals.

TABLE 4.38: Analysis of variance in role expectation of spouses, role perception and performance of farm women in health care segment among different socio-economic categories

Source of variation	d.f.	Mean Squares		
		Expectation	Perception	Performance
Between groups	4	80.291 (5.432**)	26.404 (3.480**)	4.870 (0.444)
Within groups (Error)	115	14.803	8.663	10.955

Note: Calculated 'F' values shown in parentheses.
** Significant at 1 per cent of significance.

4.4.12 *Role Performance : Health care*

Table 4.38 further points out that the computed 'F' value for role performance of farm women for the health care activities of dairy cattle and buffaloes, among different categories of socio-economic status, was not significantly different. Hence, it could be concluded that there was no difference in the performance of health care activities of dairy animals among the farm women from different socio-economic strata.

The critical difference was computed in order to find out the difference in averages in role expectation of spouses and role perception of farm women in health care segment among different categories of socio-economic status.

Health care is one of the most important activities for the productive viability of dairy animals. Table 4.39 shows that though the highest value of role expectation of spouses in this segment was observed in upper-middle category, followed by upper and middle categories of socio-economic status, but all these three categories did not differ significantly ($P>0.05$) from each other. Similarly, lower and lower-middle categories also did not differ significantly ($P>0.05$) from each other. It is obvious from Table 4.39 that middle, upper-middle and upper categories from one cluster and lower and lower-middle from another cluster which did not differ between themselves. It is, however, obvious that different categories of socio-economic status in one cluster showed significantly ($P<0.05$) higher values than the categories in the second cluster except middle and lower-middle categories between which there was no significant difference.

TABLE 4.39: Critical difference in role expectation of spouses, role perception and performance of farm women in health care segment among different socio-economic categories

	Categories of socio-economic status				
	Lower	Lower-middle	Middle	Upper-middle	Upper
Role expectation					
Average	24.60^{c}	26.00^{bc}	30.50^{ab}	32.00^{a}	30.73^{a}
S.E.	±0.59	±1.16	±2.72	±0.49	±1.72
Role perception					
Average	23.50^{ab}	23.04^{b}	23.21^{b}	25.45^{a}	26.60^{ab}
S.E.	±1.32	±0.89	±0.45	±0.38	±2.08
Role performance					
Average	—	—	—	—	—
S.E.	—	—	—	—	—

Note: The averages bearing different superscripts are significantly different at 5 per cent level of significance.

Table 4.39 indicates that farm women belonging to the upper-middle category showed significantly ($P<0.05$) higher perception of their role in this segment than in lower-middle and middle categories. But this category did not differ significantly from lower and upper categories of socio-economic status. The farm women in lower, lower-middle, middle and upper categories of socio-economic status had the role perception of the same order.

The farm women belonging to high socio-economic status perceived their role in health care of dairy animals at a higher level than their counterparts belonging to the lower categories. This might be due to the fact that farm women in higher socio-economic strata may be more aware about the importance of health care in productive animals as compared to farm women of lower socio-economic status.

4.4.13 *Role Expectation : Milk Products Making*

Table 4.40 shows that the computed 'F' value for the role expectation of spouses in the segment of milk products making among different categories of socio-economic status, was not significant. Therefore, it could be concluded that the spouses from different socio-economic categories did not statistically differ in their role expectations about this segment.

TABLE 4.40: Analysis of variance in role expectation of spouses, role perception and performance of farm women in milk products making segment among different socio-economic categories

Source of variation	d.f.	Mean Squares		
		Expectation	Perception	Performance
Between groups	4	103.671 (1.446)	2.878 (0.509)	33.798 (2.987*)
Within groups (Error)	115	71.656	5.653	11.772

Note: Calculated 'F' values shown in parentheses.
** Significant at 5 per cent level of significance.

4.4.14 *Role Perception : Milk Products Making*

The perusal of Table 4.40 further shows that the computed 'F' value for role perception of farm women in this segment also was not significant. It could, thus be concluded that there was no difference in role perception of farm women among different categories of socio-economic status with respect to the role segment of milk products making.

4.4.15 *Role Performance : Milk Products Making*

Table 4.40 further indicates that the computed 'F' values for role performance of farm women in the segment of milk products making was statistically significant ($P<0.05$). Therefore, it could be concluded that there was a significant difference in role performance of farm women in this segment among the farm women belonging to different socio-economic strata.

The critical difference was calculated in order to find out the variations in the averages of role performance of farm women in milk products making among different categories of socio-economic status.

TABLE 4.41: Critical difference in role performane of farm women in milk products making among different categories of socio-economic status

	Categories of socio-economic status				
	Lower	Lower-middle	Middle	Upper-middle	Upper
Role expectation					
Average	–	–	–	–	–
S.E.	–	–	–	–	–
Role perception					
Average	–	–	–	–	–
S.E.	–	–	–	–	–
Role performance					
Average	27.75[b]	29.00[ab]	29.50[ab]	29.55[ab]	30.82[a]
S.E.	±1.03	±0.44	±1.53	±2.43	±0.52

Note: The averages bearing different superscripts are significantly different at 5 per cent level of significance.

Table 4.41 points out that the average role performance value in lower category was significantly smaller than the role performance value of farm women in upper category. All the remaining possible pairs of socio-economic status categories did not show any significant variations in the role performance in this segment. This difference in the performance might be due to the difference in the availability of milk to the two socio-economic categories.

4.4.16 *Role Expectation : Green Fodder Production*

Perusal of Table 4.42 indicates that computed 'F' value for role expectation of spouses from farm women in green fodder production among different socio-economic categories was highly significant ($P<0.01$), concluding that there was a significant difference in role expectation of spouses among different categories of socio-economic status in the role segment of green fodder production.

4.4.17 *Role Perception : Green Fodder Production*

The computed 'F' value in analysis of variance for role perception of farm women with regard to the activities related to green fodder production was found to be highly significant ($P<0.01$). Therefore, it could be concluded that there was a statistically significant difference in the perception of farm women from different socio-economic strata about their role in activities associated with green fodder production.

TABLE 4.42: Analysis of variance in role expectation of spouses, role perception and performance of farm women in green fodder production segment among different socio-economic categories

Source of variation	d.f.	Mean Squares		
		Expectation	Perception	Performance
Between groups	4	11.625 (5.980**)	78.689 (3.491**)	88.387 (2.532*)
Within groups (Error)	115	19.736	55.805	47.033

Note: Calculated 'F' values shown in parentheses.
** Significant at 1 per cent level of significance.
* Significant at 5 per cent level of significance.

4.4.18 *Role Performance : Green Fodder Production*

Perusal of Table 4.42 further shows that the computed 'F' value for role performance of farm women in the segment of green fodder production among the farm women of different categories of socio-economic status, was statistically significant ($P<0.05$). Therefore, it could be concluded that there was a significant difference in role performance of farm women in activities associated with green fodder production among the farm women of different categories of socio-economic status.

The critical difference was calculated in order to find out the variation in the averages in role expectation of spouses, role perception and performance of farm women in relation to the segment of green fodder production among different categories of socio-economic status.

A perusal of Table 4.43 shows that the lower, lower-middle and middle categories of socio-economic status had significantly ($P<0.05$) higher values of role expectation than the upper category of socio-economic status. Table 4.43 further reveals that lower, lower-middle, middle and upper-middle categories of socio-economic status did not differ significantly ($P>0.05$) from each other. Same was the case between upper middle and upper categories of socio-economic status.

TABLE 4.43: Critical difference in role expectation of spouses, role perception and performance of farm women in green fodder production among different socio-economic categories

	Categories of socio-economic status				
	Lower	Lower-middle	Middle	Upper-middle	Upper
Role expectation					
Average	31.00^{a}	29.00^{a}	28.58^{a}	28.39^{ab}	22.45^{b}
S.E.	±1.98	±0.68	±1.34	±3.14	±0.57
Role perception					
Average	27.00^{ab}	34.50^{a}	24.04^{b}	22.00^{b}	24.74^{b}
S.E.	±5.28	±0.96	±1.15	±2.25	±3.34
Role performance					
Average	31.00^{a}	24.75^{ab}	23.60^{ab}	20.91^{b}	20.08^{b}
S.E.	±4.85	±2.07	±3.07	±1.06	±0.88

Note: The averages bearing different superscripts are significantly different at 5 per cent level of significance.

The results thus indicated that spouses in lower socio-economic category were having high role expectation from the farm women regarding the activities associated with green fodder production. The possible reason for this may be because the farm families falling in the higher socio-economic strata are probably able to hire labour which the families from lower socio-economic strata are not able to do because of obvious economic reasons and have, therefore, to depend more on family labour, including the farm women. This will result in higher expectations of the spouses for the participation of the farm women in the green fodder production related tasks.

Average role perception score of farm women in green fodder production was highest in lower-middle category and it differed significantly ($P<0.05$) from middle, upper-middle and upper categories of socio-economic status. The farm women belonging to the lower socio-economic category did not differ significantly from the farm women belonging to any other socio-economic category (Table 4.43).

The data indicated that role of farm women in green fodder production was perceived higher in lower socio-economic strata. This difference may be because of the absence of the limitations imposed by social and family norms on working out on the fields in case of the farm women belonging to the lower socio-economic strata because of economic compulsions.

The data in Table 4.43 also reveals that role performance in this segment was highest in lower category of socio-economic status and was significantly ($P<0.05$) higher than upper and upper-middle categories of socio-economic status. The role performance of farm women in lower-middle and middle categories of socio-economic status were statistically of the same order. The results indicate that the farm women in lower category of socio-economic status, participate more in activities associated with green fodder production as compared to higher categories of socio-economic strata.

It appears that the expectations of the spouses and their own perceptions about their participation in this role segment together with economic necessity and absence of social limitations results in higher performance of the farm women belonging to the lower socio-economic strata in the tasks related to green fodder production.

4.4.19 *Role Expectation : Marketing*

Table 4.44 points out that computed 'F' value for role expectation of spouses from different categories of socio-economic status, was not significant for the marketing segment. Therefore, it could be concluded that there was no difference in role expectation of spouses from different socio-economic strata in this segment.

4.4.20 *Role Perception : Marketing*

Perusal of Table 4.44 further shows that the computed 'F' value for role perception of farm women in marketing segment among different categories of socio-economic status, was also not significant. It can, therefore, be concluded that there was no significant difference in role perception of farm women in this segment among different categories of socio-economic status.

4.4.21 *Role Performance : Marketing*

Perusal of Table 4.44 also shows that the computed 'F' value for rolc performance of farm women in marketing segment was also not significant. Hence it can be concluded that there was no significant difference in role performance of farm women among different categories of socio-economic status.

TABLE 4.44: Analysis of variance in role expectation of spouses, role perception and performance of farm women in marketing segment among different socio-economic categories

Source of variation	d.f.	Mean Squares		
		Expectation	Perception	Performance
Between groups	4	18.978 (1.550)	12.472 (1.884)	6.411 (1.983)
Within groups (Error)	115	12.243	6.617	3.233

Note: Calculated 'F' values shown in parentheses.

Table 4.45 shows a profile of the livestock owners belonging to different socio-economic categories in terms of their expectations from the farm women about the activities of different segments of dairy animal production. The table also gives a profile of the perceptions and performance of the farm women belonging to different socio-economic strata, in those segments. The following significant points emerged from the data presented in Table 4.45.

1. In the lower, lower-middle and middle socio-economic strata, the expectations of the farmers from the farm women performing the feeding tasks of dairy animals, were higher than the farmers in upper-middle and upper socio-economic categories.
2. Farm women belonging to lower, lower-middle and middle socio-economic performing the feeding tasks at a significantly higher level than the farm women from the upper socio-economic category.
3. In lower-middle and middle socio-economic strata, the expectations of farmers from the farm women performing the tasks associated with breeding of dairy animals were found significantly lower than the farmers in upper-middle category.
4. Farm women from the lower and lower-middle socio-economic strata perceived their role in breeding tasks of dairy animals at a lower level than the farm women belonging to upper strata. Similarly, the perceptions in the lower-middle category were also found to be lower than in upper-middle category.
5. The expectations of the farmers from the farm women in the management tasks of dairy animals were significantly higher in lower-middle category than in upper-middle and upper categories of socio-economic status.

TABLE 4.45: Profile of role expectation of spouses and role perception and performance of farm women belonging to different socio-economic categories in different segments of dairy animal production

Segments	Pairs of socio-economic categories									
	LO v/s LM	LO v/s MD	LO v/s UM	LO v/s UP	LM v/s MD	LM v/s UM	LM v/s UP	MD v/s UM	MD v/s UP	UM v/s UP
Feeding										
Expectation	NS	NS	**High**	**High**	NS	**High**	**High**	**High**	**High**	NS
Perception	NS	NS	NS	NS	NS	NS	NS	NS	NS	NS
Performance	NS	NS	**High**	NS	NS	**High**	NS	**High**	NS	NS
Breeding										
Expectation	NS	NS	NS	NS	NS	**Low**	NS	**Low**	NS	NS
Perception	NS	NS	NS	**Low**	NS	**Low**	**Low**	NS	NS	NS
Performance	NS	NS	NS	NS	NS	NS	NS	NS	NS	NS
Management										
Expectation	NS	NS	NS	NS	NS	**High**	**High**	NS	NS	NS
Perception	NS	NS	NS	NS	NS	NS	NS	NS	NS	NS
Performance	NS	NS	NS	NS	NS	NS	NS	NS	NS	NS
Health care										
Expectation	NS	**Low**	**Low**	**Low**	NS	**Low**	**Low**	NS	NS	NS
Perception	NS	NS	NS	NS	NS	**Low**	NS	**Low**	NS	NS
Performance	NS	NS	NS	NS	NS	NS	NS	NS	NS	NS
Milk products making										
Expectation	NS	NS	NS	NS	NS	NS	NS	NS	NS	NS
Perception	NS	NS	NS	NS	NS	NS	NS	NS	NS	NS
Performance	NS	NS	NS	**Low**	NS	NS	NS	NS	NS	NS
Green fodder production										
Expectation	NS	NS	NS	**High**	NS	NS	**High**	NS	**High**	NS
Perception	NS	NS	NS	NS	**High**	**High**	**High**	NS	NS	NS
Performance	NS	NS	**High**	**High**	NS	NS	NS	NS	NS	NS
Marketing										
Expectation	NS	NS	NS	NS	NS	NS	NS	NS	NS	NS
Perception	NS	NS	NS	NS	NS	NS	NS	NS	NS	NS
Performance	NS	NS	NS	NS	NS	NS	NS	NS	NS	NS

LO = Lower; LM = Lower-middle; MD = Middle; UM = Upper-middle; UP = Upper; NS = Not significant.

6. Lower expectation of the farmers from farm women, was observed in lower socio-economic category than in the middle, upper-middle and upper categories in health care activities of dairy animals. Significantly lower expectations were also found in lower-middle category than in upper-middle and upper strata.

7. Significantly lower perceptions of farm women were found in lower-middle and middle categories than in upper-middle socio-economic category in health care activities of dairy animals.

8. The farm women in lower socio-economic category were performing the milk products making tasks at a significantly lower than the farm women in the upper category.

9. Significantly higher expectations of farmers from the farm women were found in lower, lower-middle and middle categories than in upper category of socio-economic status, in green fodder production activities.

10. As compared to middle, upper-middle and upper categories, the farm women belonging to the lower-middle category of socio-economic strata had a significantly higher perception about their role in green fodder production activities.

11. The farm women from the lower socio-economic category were performing the green fodder production activities at a significantly higher level than the farm women in upper-middle and upper socio-economic categories.

4.5 Zero Order Correlation Analysis of Independent Variables with Dependent Variables

This section deals with the nature of relationship of the studied independent variables with the dependent variables, productivity of milk and role performance of farm women in different segments of cattle and buffalo production.

In order to study the relationship between the dependent and independent variables, the values of correlation coefficients (r) were computed and tested for their statistical significance.

4.5.1 *Productivity of Milk*

Table 4.46 reveals that socio-economic status was found to be positively and significantly ($P<0.01$) associated with the productivity of milk, indicating that higher the socio-economic status, higher is the productivity of milk, while dairy herd size was found to be negatively and significantly ($P<0.05$) associated with the productivity of milk (r=–0.185) indicating that productivity of milk increased with decrease in dairy herd size. Age, position of respondents, caste, family education status, family type, land-holding, information source utilization and economic motivation were not found to be associated with the productivity of milk. This indicates that the personal characteristics of farm women did not have a major contribution in productivity of milk. However, family characteristics like socio-economic status and number of dairy animals owned by the family did contribute to the productivity of the dairy animals.

Economic motivation was found to be significantly correlated with productivity of milk by Sagar, *et al.* (1988), but the respondents in their study were farmers. In the present study a positive but non-significant relationship has been found indicating that probably the farm women were not as economically motivated as farmers.

Dairy herd size was found to be negatively associated with productivity of milk in the present study, while Sagar, *et al.* (1988), had observed a positive relationship. The observations made during the course of this study revealed that probably when dairy herd size is small i.e. one or two cows and/or buffaloes, the family can take better care of the animals. With the increase in the size of the herd, it appears that per animal care and attention decreases resulting in decrease in productivity although total milk production from the herd may be higher because of larger number of dairy animals.

TABLE 4.46: Zero order correlation coefficients between independent variables and productivity of milk

(N = 120)

	Independent variables	Productivity of milk (r)
X_1	Age	–0.078
X_2	Position of respondent	–0.002
X_3	Caste	0.091
X_4	Family type	0.093
X_5	Land-holding	0.068
X_6	Socio-economic status	0.235**
X_7	Family education status	0.152
X_8	Dairy herd size	–0.185*
X_9	Information sources utilization	0.041
X_{10}	Economic motivation	0.013

Table value of r = 0.174 and 0.228 at 5 per cent and 1 per cent of significance, respectively.

* Significant at 5 per cent level of significance.

** Significant at 1 per cent level of significance.

The socio-economic status was found to be positively correlated with productivity of milk. Obviously, livestock owners belonging to higher economic strata have better resources to meet the essential inputs requirements of feeding and health care which in turn increases the productivity of milk per animal. Moreover, they are, probably, also able to keep animals with higher genetic potential for milk production.

Table 4.47 shows that decision-making pattern in feeding was found to be positively and significantly ($P<0.05$) associated with the productivity of milk ($r = 0.201$), indicating that higher involvement of farm women in decision-making about feeding of the dairy animals will result in increased productivity of milk. Decision-making pattern in other segments was found non-significantly associated with productivity of milk.

Involvement of farm women in overall decision-making pattern of dairy animal production does not seem to be related with productivity of milk. Obviously, it is not important as to who takes the decisions, the men or the women. What is probably important is that the activities related to different segments are carried out. The significant positive relationship found in case of involvement of women in decision-making process concerning feeding activities of dairy animals with productivity of milk, however, may be because of the fact that women are generally available at home and thus have to look after the feeding of dairy animals and take decisions about it, affecting the productivity of milk. It has already been reported in this study that 35.39 per cent (overall average) of farm women were reported to be taking independent decisions and another 22.18 per cent jointly with their spouses in the matters of feeding the dairy animals (Table 4.16).

TABLE 4.47: Zero order correlation coefficients between decision-making pattern and productivity of milk

(*N = 120*)

Independent variables	Productivity of milk (r)
X_{11} Decision-making pattern:	
i. Feeding	0.201*
ii. Breeding	0.059
iii. Management	0.009
iv. Health care	0.037
v. Milk products making	0.078
vi. Green fodder production	0.069
vii. Marketing	0.042

Note: Significant at 5 per cent level of significance.

Involvement of farm women in decision-making in breeding, management, health care, milk products making, green fodder production and marketing segments was not found to be significantly correlated with productivity of milk.

Table 4.48 indicates that no significant relationship was found between the independent variable skill and the dependent variable productivity of milk.

TABLE 4.48: Zero order correlation coefficients between the independent variables (knowledge and skill) and dependent variable: productivity of milk

(*N = 120*)

Independent variables	Productivity of milk (r)
X_{14} Knowledge:	
i. Feeding	0.099
ii. Breeding	0.043
iii. Management	0.046
iv. Health care	0.003
v. Green fodder production	0.018
vi. Marketing	0.019
X_{15} Skill:	
1. Preparing and feeding of home made concentrate mixture	0.035

Note: No calculation of zero order correlation coefficient was done for knowledge of milk products making with productivity of milk, as it was not expected to have any direct relationship with productivity of milk.

4.5.2 *Role Performance*

The perusal of Table 4.49 shows that position, caste, family type, land-holding (P<0.01), age and socio-economic status (P<0.05) of the respondents were found negatively and significantly associated; whereas economic motivation was positively and significantly (P<0.05) associated with the performance of farm women in the role segment of feeding the dairy animals.

The findings indicate that level of performance of farm women in feeding the dairy animals was at a higher level by the women who were younger in age, lower in position (i.e. housewife/daughter-in-law), belonged to lower caste, nuclear families, low socio-economic status and possessed smaller land-holdings. On the other hand the older women, mothers-in-law, high caste women, having large land-holdings and belonging to higher socio-economic status, performed their role in feeding segment at a lower level.

Kaur (1981) had also reported that daughters-in-law participated more in dairy activities than the mothers-in-law. Devi (1983) had reported that caste and education were negatively associated with role performance of farm women in farm related activities.

The positive and significant relationship was found between the socio-economic status, family education status ($P<0.01$) and economic motivation ($P<0.05$) and the role performance of farm women in activities associated with breeding of dairy animals, indicating that higher the socio-economic status, family education status and economic motivation of the farm women, higher was the performance of farm women in the segment of dairy animal breeding activities.

Activities associated with dairy animal breeding were performed by the farm women whose families were socio-economically motivated. This indicated that only in more progressive families the women were performing this role which may be due to the reason that they may be more aware of the economic importance of the genetic potential of the animals and their attending breeding practices in augmenting milk production.

Table 4.49 further reveals that the position of the respondents and the dairy herd size were negatively and significantly ($P < 0.05$) associated with the role performance in managemental activities of dairy animals indicating that lower the position of respondent and smaller the dairy herd size, more will be the role performance of farm women in the management of dairy animals.

Socio-economic status was found to be positively and significantly ($P<0.01$) associated with the role performance in health care of dairy animals indicating that higher the socio-economic status, more will be the participation of the farm women in the performance of the health care activities of the dairy animals.

As indicated in Table 4.49, caste ($P < 0.05$) socio-economic status and family education status ($P < 0.01$) were found to be positively and significantly associated with role performance in the segment of milk products making, indicating that farm women belonging to higher caste, higher socio-economic status and families having higher education status were more involved in performing the activities associated with milk products making. Probably the reason is that only the economically well off families were having surplus milk for making the product and thus were involved in making milk products. The poor families were either selling or consuming milk as such and did not have any surplus milk for making its products.

TABLE 4.49: Zero order correlation coefficients between independent variables and role performance

(*N = 120*)

Independent variables	Role performance in different segments (r)						
	Feeding	Breeding	Management	Health care	Milk products making	Green fodder production	Marketing
X_1 Age	–0.178*	0.120	–0.58	0.087	–0.087	0.091	0.034
X_2 Position of respondents	–0.272**	–0.079	–0.179*	–0.020	–0.021	–0.052	–0.171
X_3 Caste	–0.214**	–0.030	–0.106	–0.101	0.194*	–0.195*	–0.029
X_4 Family type	–0.270**	0.019	–0.143	–0.054	–0.035	–0.066	–0.037
X_5 Land-holding	–0.290**	0.153	–0.083	0.001	0.033	–0.231**	0.088
X_6 Socio-economic status	–0.199*	0.238**	–0.005	0.241**	0.202**	–0.067	–0.130
X_7 Family education status	–0.067	0.254***	–0.097	0.053	0.214**	–0.016	0.138
X_8 Dairy herd size	–0.004	0.013	–0.198*	0.135	0.110	–0.331**	0.092
X_9 Information sources utilization	0.118	0.142	0.059	0.001	0.056	0.035	0.053
X_{10} Economic motivation	0.195*	0.176*	0.058	0.017	0.091	0.104	0.040

* Significant at 5 per cent level of significance.

** Significant at 1 per cent level of significance.

TABLE 4.50: Zero order correlation coefficients between decision-making pattern and role performance

(*N = 120*)

Independent variables	Role performance in different segments (r)						
	Feeding	Breeding	Management	Health care	Milk products making	Green fodder production	Marketing
X_{11} Decision-making pattern							
i. Feeding	0.235**						
ii. Breeding		0.146					
iii. Management			0.282**				
iv. Health care				0.156			
v. Milk products making					0.232**		
vi. Green fodder production						0.131	
vii Marketing							0.063

** Significant at 1 per cent level of significance.

TABLE 4.51: Zero order correlation coefficients between role expectation and role performance

(*N* = 120)

Independent variables	Role performance in different segments (r)						
	Feeding	Breeding	Management	Health care	Milk products making	Green fodder production	Marketing
X_{12} Role expectation of spouse:							
i. Feeding	0.482**						
ii. Breeding		0.158					
iii. Management			0.253**				
iv. Health care				0.089			
v. Milk products making					0.218*		
vi. Green fodder production						0.116	
vii. Marketing							0.122

* Significant at 5 per cent level of significance.
** Significant at 1 per cent level of significance.

TABLE 4.52: Zero order correlation coefficients between role perception and role performance of farm women

(*N = 120*)

Independent variables	Role performance in different segments (r)						
	Feeding	Breeding	Management	Health care	Milk products making	Green fodder production	Marketing
X_{13} Role perception of farm women:							
i. Feeding	0.621**						
ii. Breeding		0.121					
iii. Management			0.382**				
iv. Health care				0.159			
v. Milk products making					0.775**		
vi. Green fodder production						0.113	
vii. Marketing							0.116

** Significant at 1 per cent level of significance.

Caste ($P < 0.05$), land-holding and dairy herd size ($P < 0.01$) were found to be associated with the performance of the activities related to green fodder production, indicating that lower the caste, smaller the land-holding and herd size, higher was the performance of farm women in green fodder production activities. Other variables were found to be non-significantly associated with role performance in green fodder production.

It is obvious from the findings the farm women belonging to marginal and small farmers' categories having limited land and one or two dairy animals, participate in the activities related with the production of green fodder at a higher level, may be only on a small piece of land diverted from food crops to fodder cultivation. It appears that due to cultural restrictions the contribution of farm women belonging to higher caste in fodder cultivation related activities is at a level lower than the farm women belonging to the lower caste.

A non-significant relationship was found between the role performance in marketing activities and all the studied independent variables. Obviously, the farm women were not involved in marketing activities, because most of these activities are performed by livestock owners. This resulted in non-significant relationship between role performance in marketing activities and the independent variables taken under the study.

The perusal of Table 4.50 shows that the decision-making pattern in feeding, management and milk products making were positively and significantly ($P < 0.01$) correlated with the role performance in the respective segments indicating that an increase in the involvement of farm women in decision-making process related to the activities associated with these segments leads to their increased performance in these segments. Decision-making pattern in other segments of dairy animal production was not found to be associated with the performance of the farm women in the respective segments.

Table 4.51 indicates that the role expectation of spouses in activities related to feeding, management ($P < 0.01$) and milk products making ($P < 0.05$) segments were found to be positively and significantly associated with the role performance of farm women in the respective segments, indicating that higher the role expectation of the spouses, higher was the role performance of farm women in these segments. The positive and significant correlation between the expectations of the spouses and the performance of the farm women, in these role segments indicates that farm women were performing their role in these segments as per the expectations of their spouses and that there was no conflict between the farm women and their spouses in these segments. The relationship was not found significant in other role segments.

Table 4.52 points out that the role perception of farm women in the segment of feeding, management and milk products making was found to be positively and significantly ($P < 0.01$) associated with their performance in the respective segments indicating in that in these segments the farm women were performing their role as they themselves had perceived it. Their performance of the activities in the segments of feeding, management and milk products making was in consonance with their perception about their role in these segments.

The results shows that the expectations of the spouses and the perceptions of the farm women were significantly correlated with the performance of farm women in the role segments of feeding, management and milk products making. This indicated that there was no conflict between the spouses and the farm women in the role expectation, perception and performance in these role segments.

Table 4.53 indicates that knowledge level of the farm women about feeding of the dairy animal was found to be positively and significantly ($P < 0.05$) associated with their role performance in this segment indicating that higher the level of knowledge of farm women about feeding of dairy animals, higher will be their performance in this segment. Knowledge level of no other segment was found to be significantly correlated with the performance of farm women in this segment.

Table 4.53 further shows that the level of skill possessed by farm women in clean milk production was positively and significantly ($P<0.05$) associated with their role performance in management segment indicating that as the skill in clean milk production increases, the performance of farm women in management of dairy cattle and buffaloes also improves. Their level of skill in the preparation and feeding of home-made concentrate mixture was not found to be significantly correlated with their performance in feeding segment.

On the basis of the findings in the preceding pages the following important points emerged.

1. Families belonging to higher socio-economic strata and having smaller number of dairy animals achieved a higher productivity of milk from their dairy animals.

TABLE 4.53: Zero order correlation coefficients between knowledge and skill levels of farm women in different segments and role performance (*N = 120*)

Independent variables	Role performance in different segments (r)						
	Feeding	Breeding	Management	Health care	Milk products making	Green fodder production	Marketing
X_{14} Knowledge:							
i. Feeding	0.235**						
ii. Breeding		0.176					
iii. Management			0.190				
iv. Health care				0.114			
v. Milk products making					0.033		
vi. Green fodder production						0.155	
vii. Marketing							0.114
X_{15} Skill:							
i. Clean milk production			0.199*				
ii. Preparing and feeding home-made concentrate mixture	0.121						

* Significant at 5 per cent level of significance.
** Significant at 1 per cent level of significance.

2. Higher level of participation of farm women in the decision-making about feeding of dairy animals positively influenced the productivity of milk in dairy animals.
3. Farm women belonging to younger age group, holding lower family position (daughters and daughters-in-law)), belonging to lower castes, nuclear families, lower socio-economic strata, possessing smaller land-holding and higher economic motivation were performing the tasks associated with feeding of the dairy animals at a higher level.
4. Tasks associated with the breeding activities of dairy animals were performed at a higher level by the farm women belonging to higher socio-economic strata and who had a higher family education status and economic motivation.
5. In the families which had smaller number of dairy animals, the daughters and daughters-in-law were performing the managemental tasks of dairy animals at a higher level.
6. Farm women belonging to the families in higher socio-economic strata were performing the health care tasks of dairy animals at a higher level.
7. The milk products making tasks were performed at a higher level by the farm women belonging to higher castes, higher socio-economic status and having a higher family education status.
8. Farm women belonging to lower castes, having small land-holdings and fewer dairy animals were performing the tasks related to green fodder production at a higher level.
9. Farm women who were participating to decision-making at a higher level in activities related to feeding, management and milk products making segments of dairy animal production were also performing their role in these segments at a higher level.
10. Farm women were performing the tasks in feeding, management and milk products making segments of dairy animal production as per the expectations of the spouses in terms of who should do those tasks.
11. Farm women were performing the tasks related to the feeding, management and milk products making segments of dairy animal production as per their own perceptions about who should perform those tasks.
12. There did not appear to be any conflict between the livestock

owners and the farm women as far as the feeding, management and milk products making segments of dairy animal production were concerned.

13. Knowledge level of farm women in the feeding of dairy animals significantly affected her performance in this segment.
14. Level of skill possessed by the farm women in clean milk production significantly affected her performance in the management of dairy animals.

4.6 Multipurpose Regression Analysis of Dependent Variables with Independent Variables

The multiple regression analysis was used to determine the degree to which the dependent variables could be predicted with the help of independent variables.

4.6.1 *Productivity of Milk*

In Table 4.54 a regression model was used utilizing ten independent variables i.e. age, position, caste, family type, land-holding, socio-economic status, family education status, dairy herd size, information sources utilization and economic motivation to explain the variability of productivity of milk. The model explains about 58 per cent variability of dependent variable and was found significant ($P < 0.01$).

The combined significant explanation provided by this model does not conclude directly that each of the ten independent variables included in the model contribute significantly for explaining the variability of dependent variable. To isolate the independent variables contributing significantly in explaining the variability of dependent variable, 't' test was used, which showed that only socio-economic status was the independent variable which contributed significantly ($P<0.01$) in explaining the variability of productivity of milk.

The significant variable showed positive coefficient which concluded that increase in socio-economic status is associated with the increase in the productivity of milk.

All other remaining independent variables did not contribute significantly in influencing the productivity of milk.

TABLE 4.54: Multiple regression between independent variables and productivity of milk

(*N* = 120)

	Independent variables	b	t	R^2	F
X_1	Age	–0.3410	0.2259	0.5785	5.4336**
X_2	Position of respondent	–2.7786	0.3264		
X_3	Caste	0.1712	0.3520		
X_4	Family type	1.8048	1.8850		
X_5	Land-holding	5.4145	0.2460		
X_6	Socio-economic status	3.2516	3.5830**		
X_7	Family education status	1.3432	0.4722		
X_8	Dairy herd size	–6.3691	1.4068		
X_9	Information sources utilization	0.4768	0.5244		
X_{10}	Economic motivation	0.5354	1.2451		

Table value of *t* = 1.980 and 2.617 at 5 per cent and 1 per cent level of significance, respectively.

Table value of *F* = 1.850 and 2.360 at 5 per cent and 1 per cent of significance, respectively.

** Significant at 1 per cent level of significance.

Socio-economic status remains the major contributing variable in predicting the productivity of milk. Sagar, *et al.* (1988 and 1989) reported similar findings that the annual family income, feeding of green fodder and economic motivation were significantly related with productivity of milk. Higher socio-economic status means more availability of resources and inputs, an essential aspect in the dairy animal production.

TABLE 4.55: Multiple regression between the independent variables and role performance of farm women in the tasks related to feeding of dairy animals

(*N* = 120)

	Independent variables	b	t	R^2	F
X_1	Age	–0.1498	0.6629	0.3508	3.9861**
X_2	Position of respondent	–2.9138	1.7537		
X_3	Caste	–2.7588	3.1954**		
X_4	Family type	–0.4920	1.5258		
X_5	Land-holding	–3.6006	1.8127		
X_6	Socio-economic status	–0.4373	3.5103**		
X_7	Family education status	–0.4740	0.3485		
X_8	Dairy herd size	–1.7254	1.0449		
X_9	Information sources utilization	0.3614	0.4459		
X_{10}	Economic motivation	1.5482	3.4421**		

** Significant at 1 per cent level of significance.

4.6.2 *Role Performance : Feeding*

To predict the role performance in the segment feeding, ten independent variables i.e. age, position, caste, family type, land-holding, socio-economic status, family education status, dairy herd size, information sources utilization and economic motivation were fitted in the regression equation to explain the variability in the role performance in feeding. This model explained about 35 per cent variability of the dependent variable and was found highly significant ($P < 0.01$).

To isolate the independent variables contributing significantly in explaining the variability of dependent variable, 't' test was computed which concluded that only caste, socio-economic status and economic motivation were contributing significantly ($P < 0.01$) in explaining the variability of role performance in feeding. Caste and socio-economic status showed negative coefficients which concluded that decrease in these independent variables will lead to increase role performance in feeding the dairy cattle and buffaloes, whereas economic motivation showed positive coefficient, which indicated that farm women's performance in feeding activities of the dairy animals will be at a higher level with higher economic motivation levels in them. All other independent variables did not contribute at statistically significant level in influencing the role performance of farm women in the feeding segment.

TABLE 4.56: Multiple regression between the independent variables and role performance of farm women in the tasks related to breeding of dairy animals ($N = 120$)

	Independent variables	b	t	R^2	F
X_1	Age	0.7848	1.5670	0.2483	2.6871*
X_2	Position of respondent	–0.5736	0.9713		
X_3	Caste	–0.7511	0.9078		
X_4	Family type	–0.5038	1.4657		
X_5	Land-holding	–0.6773	1.4819		
X_6	Socio-economic status	0.7804	2.5308*		
X_7	Family education status	0.1025	0.1743		
X_8	Dairy herd size	0.3931	1.6930		
X_9	Information sources utilization	0.1216	1.1542		
X_{10}	Economic motivation	0.2451	1.8925		

* Significant at 5 per cent level of significance.

4.6.3 *Role Performance — Breeding*

Ten independent variables viz. age, position, caste, family type, land-holding, socio-economic status, family education status, dairy herd size, information sources utilization and economic motivation were used to explain the variability of dependent variable, role performance of farm women in breeding activities of dairy animals. The model explained about 25 per cent variability of the dependent variable and was found to be statistically significant ($P < 0.05$).

The 't' test concluded that only socio-economic status was the independent variable which contributed significantly ($P < 0.05$) in explaining the variability of the role performance of farm women in segment breeding activities of dairy animals. It showed positive coefficient which indicated that increase in socio-economic status leads to increase in the role performance of farm women in this segment. None of the other variables showed significant relationship with the performance of farm women in this segment.

TABLE 4.57: Multiple regression between the independent variables and the role performance of farm women in the tasks related to management of dairy animals

(*N = 120*)

	Independent variables	b	t	R^2	F
X_1	Age	–0.5453	0.8359	0.3263	3.9175**
X_2	Position of respondent	–1.2642	1.0023		
X_3	Caste	–1.1656	1.8951		
X_4	Family type	2.5630	0.8028		
X_5	Land-holding	–1.3527	0.9978		
X_6	Socio-economic status	–0.4238	1.2277		
X_7	Family education status	0.1810	0.8192		
X_8	Dairy herd size	–1.0731	4.0757**		
X_9	Information sources utilization	0.1681	0.4557		
X_{10}	Economic motivation	0.6198	1.2310		

** Significant 1 per cent level of significance.

4.6.4 *Role Performance : Management*

The variability of the role performance of farm women in the management of dairy animals was found to be 32.63 per cent on account of the independent variables age, position, caste, family type (nuclear or joint), land-holding, socio-economic status, family education status, dairy herd size and economic motivation was found to be statistically significant ($P < 0.05$).

To isolate the independent variables contributing significantly in explaining the variability of dependent variable, 't' was computed which showed that only dairy herd size was the independent variable which contributed significantly ($P < 0.01$) in explaining the variability of management of dairy cattle and buffaloes.

The significant independent variable, dairy herd size, showed negative coefficient which indicated that smaller the dairy herd size higher the performance of farm women in management tasks of dairy animals.

All other independent variables did not contributed significantly in influencing the role performance of farm women in management of dairy animals.

It can be concluded from the findings that when there is smaller dairy herd size, then it becomes easy and convenient for farm women to manage it well.

TABLE 4.58: Multiple regression between the independent variables and role performance of farm women in the tasks related to health care of dairy animals

(N = 120)

	Independent variables	b	t	R^2	F
X_1	Age	0.3957	1.5664	0.2605	2.4861**
X_2	Position of respondent	0.4718	1.1946		
X_3	Caste	0.2556	1.7496		
X_4	Family type	–0.8291	0.4770		
X_5	Land-holding	–0.6347	1.2457		
X_6	Socio-economic status	0.6095	2.3684*		
X_7	Family education status	1.1981	0.3195		
X_8	Dairy herd size	0.2210	0.9027		
X_9	Information sources utilization	0.1657	1.4931		
X_{10}	Economic motivation	0.3717	0.2198		

* Significant at 5 per cent level of significance

** Significant 1 per cent level of significance.

4.6.5 *Role Performance : Health Care*

To predict the role performance of farm women in health care activities of dairy cattle and buffaloes, ten independent variables i.e. age, position, caste, family type, land-holding, socio-economic status, family education status, dairy herd size, information sources utilisation and economic motivation were fitted in the regression equation to explain the variability of the role performance in health care of dairy animals. The model explained about 26 per cent variability of the dependent variable and was found significant ($P < 0.01$). In isolating the independent variables contributing significantly to the variability, 't' test was computed which showed that only socio-economic status was contributing significantly ($P < 0.05$) in explaining the variability of role performance of farm women in the health care segment of dairy animal production.

All other remaining independent variables did not contribute significantly in influencing the role performance of farm women in this segment.

It can be concluded that farm women in higher socio-economic status perform the role in health care of dairy animals at a higher level than the farm women belonging to lower socio-economic status. It may be because they were more sensitive to keep their animals healthy.

TABLE 4.59: Multiple regression between the independent variables and role performance of farm women in the tasks related to milk products making

(N = 120)

	Independent variables	b	t	R^2	F
X_1	Age	-0.5648	1.4447	0.2674	2.4553**
X_2	Position of respondent	0.2744	0.4406		
X_3	Caste	0.4677	1.4944		
X_4	Family type	0.8291	0.8366		
X_5	Land-holding	-0.6347	1.6750		
X_6	Socio-economic status	0.6095	2.6774**		
X_7	Family education status	0.1981	0.3195		
X_8	Dairy herd size	0.2210	0.9027		
X_9	Information sources utilization	0.1659	1.4931		
X_{10}	Economic motivation	0.3717	0.2198		

** Significant 1 per cent level of significance.

4.6.6 *Role Performance : Milk Products Making*

The regression model explains about 27 per cent variability of the dependent variable by the ten independent variables i.e. age, position of respondents, caste, family type (nuclear or joint), land-holding, socio-economic status, dairy herd size, information sources utilization and economic motivation fitted in the model and was found significant (P<0.01). The 't' test indicated that only socio-economic status contributed significantly (P<0.01) in explaining the variability of role performance in milk products making. The significant independent variable i.e. socio-economic status, showed positive coefficient, which indicated that increase in socio-economic status will lead to a higher level of role performance by farm women in the tasks related to milk products making.

All the other remaining independent variables were not found to be significantly related with the role performance in this segment.

Caste and family education which were found having significant correlation with the role performance in milk products making through zero order correlation coefficient, could not sustain their independent contribution through multiple regression analysis. These variables are important components of socio-economic status which was found to be most important variable contributing to productivity of milk. Milk products such as *Ghee, Khoa* etc. are given more importance in higher socio-economic status groups which probably make their women perform these tasks at a higher level. Moreover, families belonging to higher socio-economic status also have more surplus milk to divert for making products.

TABLE 4.60: Multiple regression between the independent variables and role performance of farm women in the tasks related to green fodder production

(N = 120)

	Independent variables	b	t	R^2	F
X_1	Age	0.1394	0.4206	0.3872	3.5952**
X_2	Position of respondent	–1.2642	1.3391		
X_3	Caste	–1.1656	2.5076*		
X_4	Family type	–0.6747	0.8028		
X_5	Land-holding	–0.4256	0.9978		
X_6	Socio-economic status	–0.4230	0.2471		
X_7	Family education status	–0.4307	0.8192		
X_8	Dairy herd size	–1.0731	2.5683*		
X_9	Information sources utilization	0.1681	0.4557		
X_{10}	Economic motivation	0.1764	1.2310		

* Significant at 5 per cent level of significance.
** Significant at 1 per cent level of significance.

4.6.7 *Role Performance : Green Fodder Production*

Table 4.60 shows ten independent variables i.e. age, position of respondent, caste, family type, land-holding, socio-economic status, family education status, dairy herd size, information sources utilization and economic motivation, which were used to explain the variability of role performance of farm women in the tasks related with green fodder production. The model explains about 39 per cent variability of dependent variable and was found highly significant ($P < 0.01$).

The 't' test concluded that caste and dairy herd size were the two independent variables which contributed significantly ($P < 0.05$) in explaining the variability of role performance of farm women in green fodder production.

Both the significant independent variables i.e. caste and dairy herd size, showed negative coefficients which indicated that the farm women belonging to lower castes or having fewer dairy animals were performing the tasks of this segment at a higher level than those belonging to upper castes or having more dairy animals. The remaining independent variables were not found significantly related with the role performance of farm women in this segment.

TABLE 4.61: Multiple regression between the independent variables and the role performance of farm women in the tasks related to marketing

(N = 120)

	Independent variables	b	t	R^2	F
X_1	Age	0.9675	1.2340	0.2021	1.7951
X_2	Position of respondent	–0.5197	0.8956		
X_3	Caste	–0.8494	1.7761		
X_4	Family type	–0.6317	0.6867		
X_5	Land-holding	–0.6994	0.1506		
X_6	Socio-economic status	–0.8666	1.0263		
X_7	Family education status	0.5320	0.9230		
X_8	Dairy herd size	–0.3740	1.6586		
X_9	Information sources utilization	0.1395	1.3449		
X_{10}	Economic motivation	0.2208	1.4027		

From the findings it can be concluded that the farm women of lower caste contribute in green fodder production at a higher level. Similarly, in the families which have smaller number of dairy animals, the farm women were involved in green fodder production task at a higher level. This was, probably because having fewer dairy animals in the family decreases the workload of farm women and hence they were able to devote more time to this segment of dairy animal production.

4.6.8 *Role Performance : Marketing*

In Table 4.61 the regression model explains about 20 per cent variability of rôle performance in the marketing segment by the ten independent variables which were fitted in the model. None was found statistically significant.

The 't' test showed that none of the independent variables contributed significantly. Similar results were observed in zero order correlation coefficients. In fact, the farm women do not perform much of the role/activities in the marketing segment of dairy animal production. Low expectation of spouses and low perception of farm women were also observed for this role segment. This phenomenon of low expectation, perception and performance of this role segment *vis-a-vis* the farm women might probably be due to the reason that marketing being an outdoor activity and involves and deals with monetary matters is, therefore, considered a job to be undertaken by the livestock owners themselves, they being the bread earners of the family.

4.7 Direct and Indirect Effects of Selected Independent Variables on Dependent Variables

The correlation coefficient (r) and multiple regression analysis indicated that the observed relationship between the independent and dependent variables is partially absolute and partially relative. A portion of observed relationship is the contribution made by other independent variables through which the independent variables exercise their influence jointly. It was, therefore, considered necessary to study the extent of direct and indirect influence of independent variables on dependent variables. Path analysis technique was, therefore, used to study these indirect relationships. According to LI (1955) path coefficient is an absolute number without any physical unit, whatever the actual units of measurement for the variables. It also shows direction.

4.7.1 *Production of Milk*

Table 4.62 shows that the maximum direct effects were exercised by socio-economic status (X_6), followed by dairy herd size (X_8) and land-holding (X_5), in that order. Socio-economic status (X_6) and land-holding (X_5) had positive direct effect; whereas dairy herd size (X_8) had negative direct effect on the productivity of milk.

Table 4.62 and Fig. 4.8 indicate that the ranks of the variables with respect to the direct effects and total indirect effects, are not the same, except socio-economic status (X_6) which occupies the first rank in both the irect and total indirect effects.

Table 4.62 further reveals that eight out of the ten independent variables had their largest indirect effects through socio-economic status (X_6) and the remaining two with dairy herd size (X_8) on the productivity of milk of dairy animals.

The path analysis has indicated that socio-economic status (X_6) is the most important variable which affects productivity of milk in dairy animals both directly and indirectly. Socio-economic status, however, has also shown it's indirect effect through dairy herd size (X_8) on productivity of milk. The results have revealed that productivity of milk by dairy animals largely depends upon the resource availability of the livestock owners. Although dairy herd size has shown a mild negative direct effect on productivity of milk, this negative direct effect if combined with the positive high direct effect of resource availability indicates that probably even the people belonging to the higher socio-economic status are not keeping large number of dairy animals but are achieving higher milk production through smaller number of dairy animals by keeping them on a better and balance feeding regime due to their higher resource availability.

In a recent and independent sample survey (N = 302) conducted by the Krishi Vigyan Kendra of Indian Veterinary Research Institute, Izatnagar* in Bareilly District, also the target population of the present research study, it was found that per hectare dairy animal holding was 2.78, 1.55, 0.90 and 0.73 for the marginal, small, medium and large farmers' categories. The dairy animal holding included desi cows and buffaloes, crossbred cows and improved buffaloes. Since land-holding is an important component in categorisation of individuals on their socio-economic status, the sample survey data support the findings of this study.

---

* Personal communication from Shri M.S. Rathore, Training Associate, (Ani. Hubs.), K.V.K., Joint Directorate of Extension Education, I.V.R.I. Izatnagar.

TABLE 4.62: Direct and indirect effects of exogenous variables on productivity of milk

(*N* = *120*)

Exogenous variables	Direct Effects		Total Indirect Effects		Largest Indirect Effects Through Single Variable	
	Effect	Rank	Effect	Rank	Effect	Variable No.
X_1 Age	–0.0571	IV	–0.0209	X	0.0379	X_8
X_2 Position of respondents	0.0399	V	–0.0275	IX	0.0203	X_6
X_3 Caste	–0.0039	X	0.0953	IV	0.2310	X_6
X_4 Family type	–0.0158	IX	0.0773	VI	0.1166	X_6
X_5 Land-holding	0.0927	III	0.1669	II	0.3058	X_6
X_6 Socio-economic status	0.4474	I	0.2823	I	0.0929	X_8
X_7 Family education status	0.0211	VIII	0.1579	III	0.3316	X_6
X_8 Dairy herd size	–0.1856	II	0.0920	V	0.0561	X_6
X_9 Information sources utilization	0.0296	VI	0.0710	VII	0.2132	X_6
X_{10} Economic motivation	0.0219	VII	0.0371	VIII	0.1424	X_6

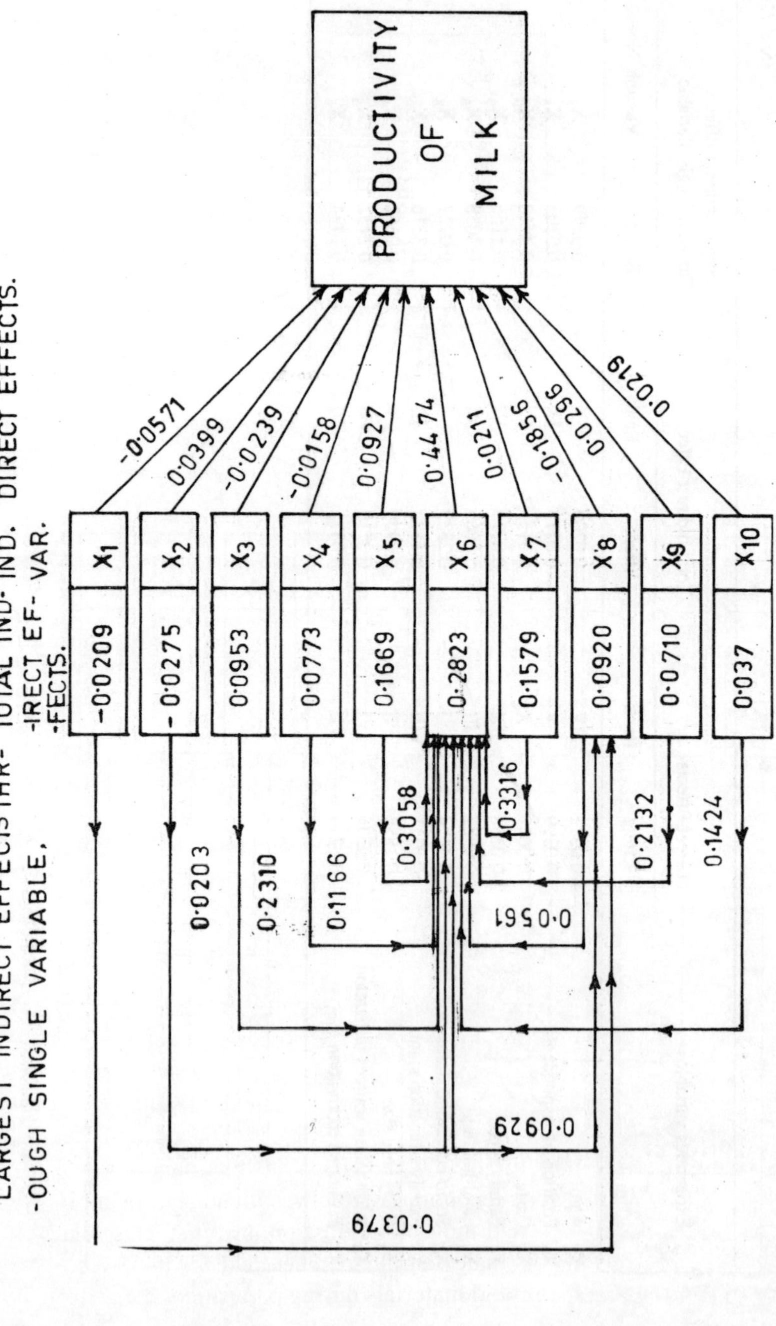

FIG 4.8: Path analysis showing direct and indirect effects of independent variables on productivity of milk

4.7.2 *Role Performance : Feeding*

Multivariate path analysis with the role performance of farm women in feeding segment as dependent variable and ten independent variables, was carried out and the results obtained are presented in Table 4.63.

Rankings of direct effect (Table 4.61) indicate that the higher direct effect was exercised by economic motivation (X_{10}), followed by land-holding (X_5) and family type (X_4), in that order. Land-holding (X_5) and family type (X_4) had negative, whereas economic motivation (X_{10}) had positive direct effect on the role performance of farm women in the tasks related to feeding the dairy animals.

Table 4.63 and Fig. 4.9 indicate that the ranks of the variables with respect to the direct effect and the total indirect effects are not the same. The maximum total indirect effect was found to be exercised by socio-economic status (X_6), position of respondent (X_2), economic motivation (X_{10}), in that order.

Table 4.63 further reveals that most of the independent variables had their largest indirect effect on the role performance of farm women in this segment through land-holding (X_5) followed by family type (X_4) and socio-economic status (X_6).

4.7.3 *Role Performance : Breeding*

Table 4.64 indicates that the highest direct effect was recorded through caste (X_3), followed by age (X_1) and socio-economic status (X_6) in that order. Age and socio-economic status had positive, whereas caste had negative direct effect on role performance of farm women in task related to breeding of dairy animals.

Table 4.64 and Fig. 4.10 indicates that the ranks of the variables with respect to the direct effect and the total indirect effects are not the same except caste (X_3) indicating that it was the most important contributing variable, both directly and indirectly, in the role performance of farm women in this segment.

Table 4.64 further points out that majority of the independent variables had their largest indirect effect through socio-economic status (X_6) on the role performance of farm women in this segment.

Farm women on the whole perform very little role in breeding. Path analysis, however, indicated that role in breeding is affected by caste and socio-economic status directly. Age is also an important variable which had direct effect. Socio-economic status was found significant in both zero order correlation and multiple regression analysis. This role, in general, is performed by male family members except in few activities like 'arranging household materials during parturition' etc.

TABLE 4.63: Direct and indirect effects of exogenous variables on the role performance of farm women in the activities/tasks related to feeding of dairy animals

($N = 120$)

Exogenous variables	Direct Effects		Total Indirect Effects		Largest Indirect Effects Through Single Variable	
	Effect	Rank	Effect	Rank	Effect	Variable No.
X_1 Age	–0.0641	VII	–0.1119	IV	–0.0379	X_3
X_2 Position of respondents	–0.0476	IX	–0.2125	II	–0.1337	X_4
X_3 Caste	–0.1409	IV	–0.0720	VI	–0.0971	X_5
X_4 Family type	–0.1966	III	0.0580	VIII	–0.0654	X_5
X_5 Land-holding	–0.2302	II	0.0583	VII	–0.0883	X_4
X_6 Socio-economic status	–0.0930	VI	–0.2806	I	0.1601	X_5
X_7 Family education status	–0.0502	VIII	0.0519	X	0.1458	X_6
X_8 Dairy herd size	–0.1012	V	0.0903	V	0.1592	X_{10}
X_9 Information sources utilization	0.0454	X	0.0580	IX	0.0454	X_6
X_{10} Economic motivation	0.3512	I	0.1618	III	0.0579	X_8

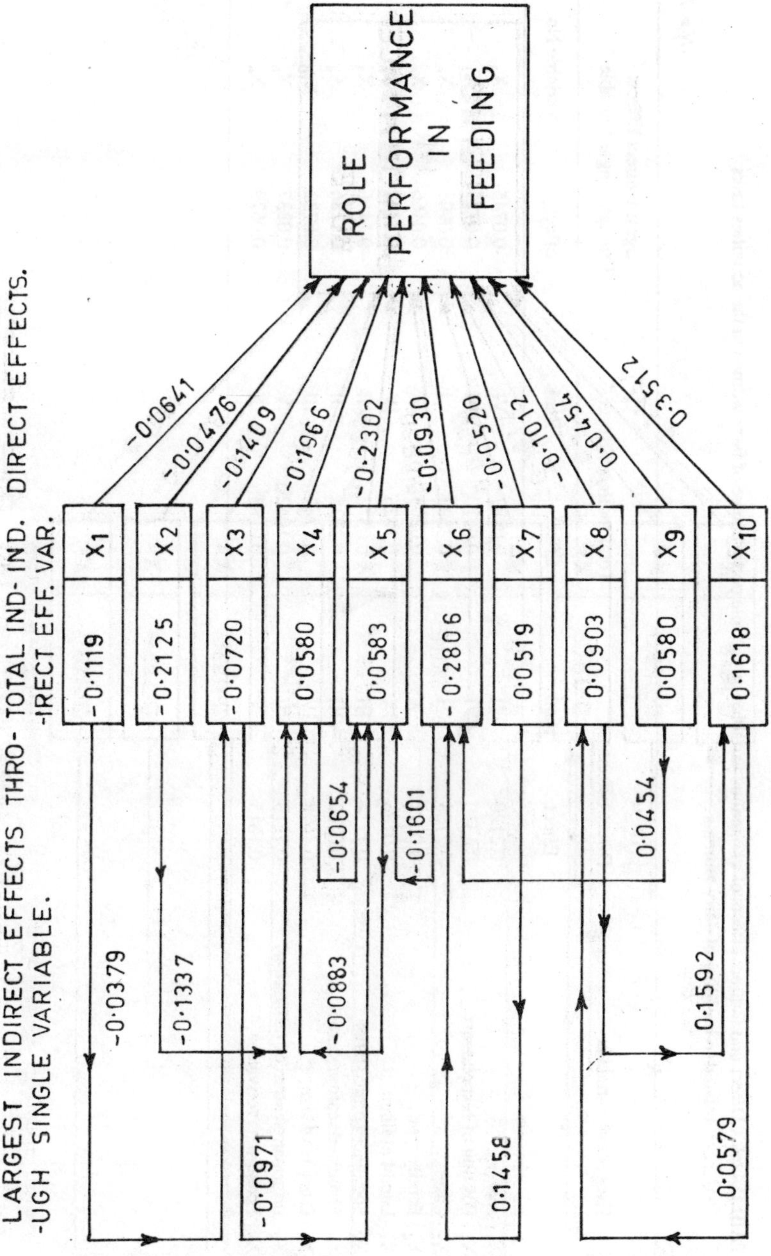

FIG 4.9: Path analysis showing direct and indirect effects of independent variables on role performance in feeding

TABLE 4.64: Direct and indirect effects of exogenous variables on the role performance of farm women in the activities/tasks related to breeding of dairy animals

(*N* = *120*)

Exogenous variables	Direct Effects		Total Indirect Effects		Largest Indirect Effects Through Single Variable	
	Effect	Rank	Effect	Rank	Effect	Variable No.
X_1 Age	0.2120	II	0.0887	VI	–0.0745	X_3
X_2 Position of respondents	0.1330	VI	0.0510	VIII	0.0729	X_1
X_3 Caste	–0.2768	I	0.2593	I	0.0901	X_6
X_4 Family type	–0.0071	X	0.0203	IX	0.0466	X_6
X_5 Land-holding	–0.0186	IX	0.1626	IV	0.1192	X_6
X_6 Socio-economic status	0.1714	III	0.0572	VII	–0.1456	X_3
X_7 Family education status	0.0260	VIII	0.2235	II	0.1286	X_6
X_8 Dairy herd size	0.1698	IV	0.1747	III	0.0732	X_{10}
X_9 Information sources utilization	0.1218	VII	0.1042	V	0.0837	X_6
X_{10} Economic motivation	0.1615	V	0.0185	X	0.0679	X_6

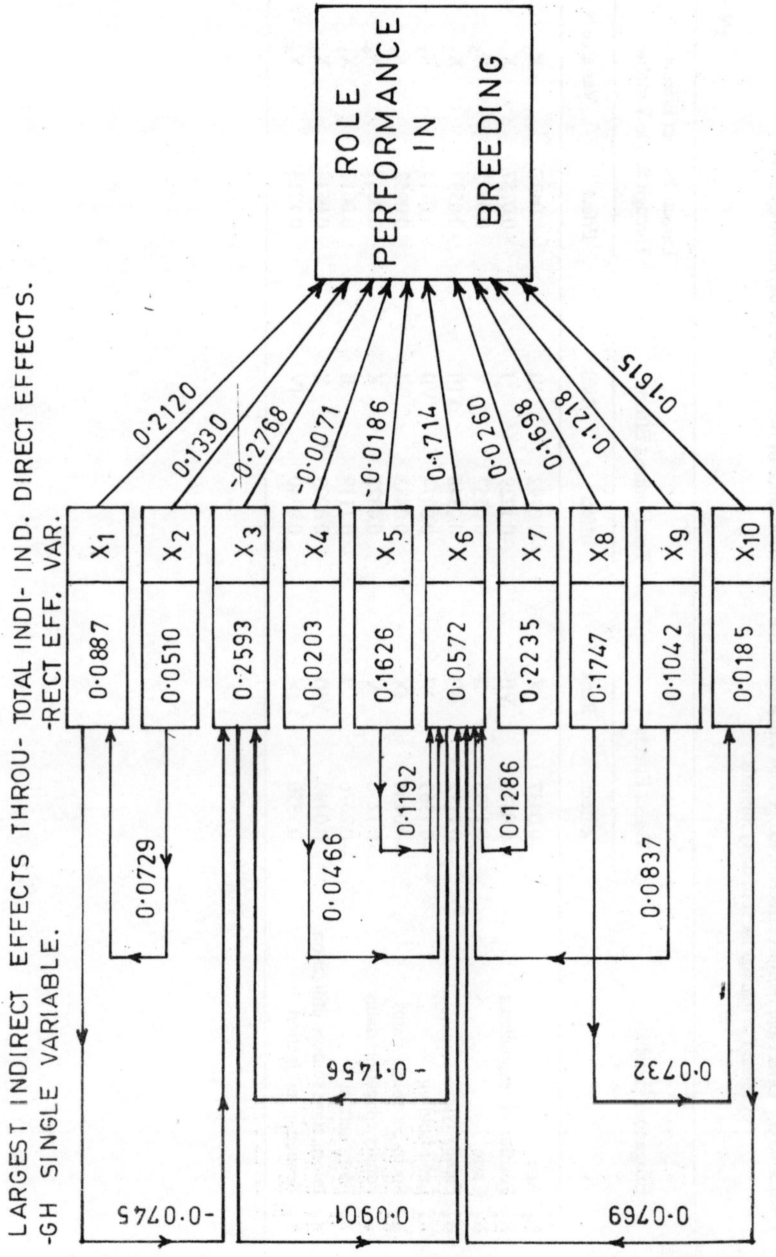

FIG 4.10: Path analysis showing direct and indirect effects of independent variables on role performance in breeding

TABLE 4.65: Direct and indirect effects of exogenous variables on the role performance of farm women in the activities/tasks related to management of dairy animals

(*N = 120*)

Exogenous variables	Direct Effects		Total Indirect Effects		Largest Indirect Effects Through Single Variable	
	Effect	Rank	Effect	Rank	Effect	Variable No.
X_1 Age	–0.0437	X	–0.0792	III	–0.0437	X_3
X_2 Position of respondents	–0.0878	VII	–0.1229	I	–0.0757	X_4
X_3 Caste	–0.1535	II	0.0557	V	0.0575	X_5
X_4 Family type	–0.1112	VI	0.0348	VIII	–0.0597	X_2
X_5 Land-holding	–0.1363	III	0.0474	VII	0.0714	X_8
X_6 Socio-economic status	–0.0484	IX	0.0483	VI	0.0952	X_3
X_7 Family education status	0.1270	V	0.0322	X	0.0687	X_6
X_8 Dairy herd size	–0.2676	I	0.1119	II	0.0612	X_5
X_9 Information sources utilization	0.0499	VIII	0.0003	X	0.0639	X_7
X_{10} Economic motivation	0.1350	IV	0.0740	IV	0.1213	X_8

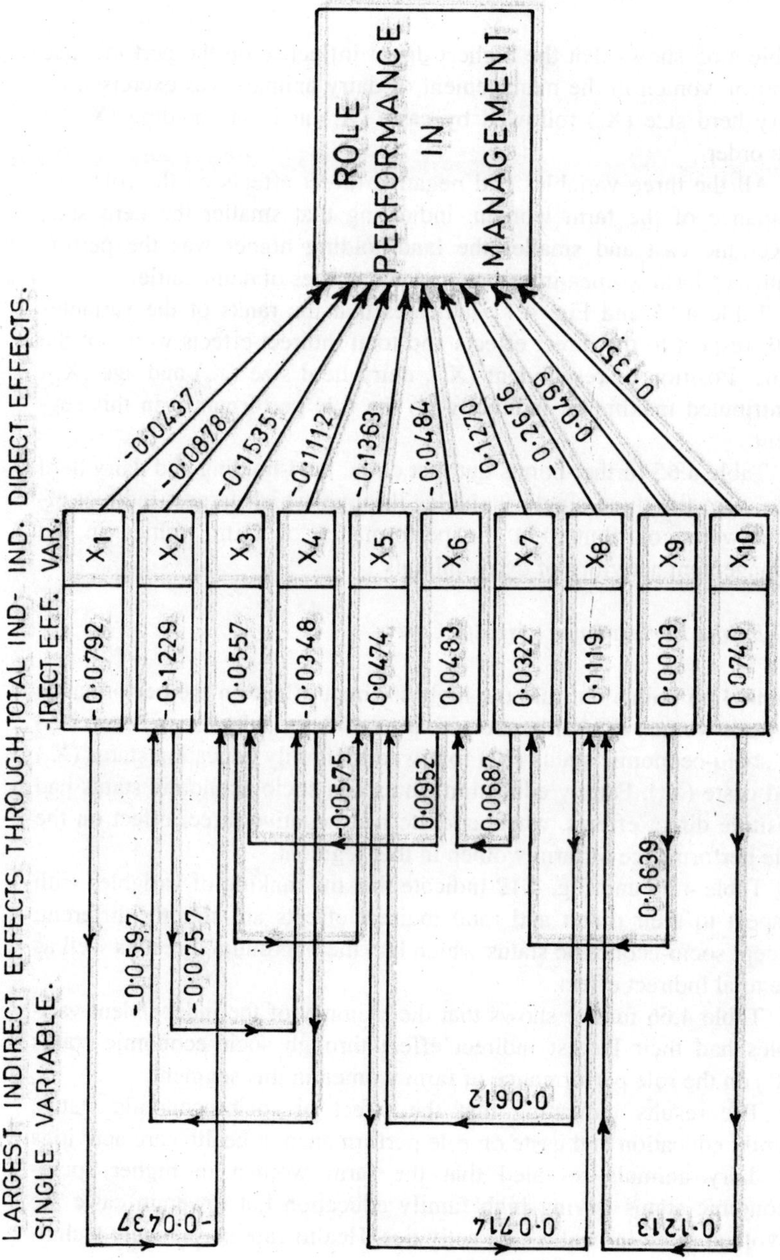

FIG 4.11: Path analysis showing direct and indirect effects of independent variables on role performance in management

4.7.4 *Role Performance : Management*

Table 4.65 shows that the highest direct influence on the performance of farm women in the management of dairy animals was exercised by dairy herd size (X_8) followed by caste (X_3) and land-holding (X_5) in that order.

All the three variables had negative direct effects on the role performance of the farm women, indicating that smaller the herd size, lower the cast and smaller the land-holding higher was the performance of farm women in management activities of dairy cattle.

Table 4.65 and Fig. 4.11 indicates that the ranks of the variables with respect to the direct effects and total indirect effects were not the same. Position of respondent (X_2), dairy herd size (X_8) and age (X_1), contributed maximum, indirectly, to the role performance in this segment.

Table 4.65 further points out that caste, land-holding and dairy herd size were the three single variables through which largest indirect effects were contributed in the performance of farm women in the management activities of dairy animals.

4.7.5 *Role Performance : Health Care*

Table 4.66 points out that the highest direct effect on the performance of farm women in health care activities of dairy animals was exercised by socio-economic status (X_6) followed by family education status (X_7) and caste (X_3). Family education status and socio-economic status had positive direct effects, whereas caste had negative direct effect on the role performance of farm women in this segment.

Table 4.66 and Fig. 4.12 indicate that the ranking of variables with respect to their direct and total indirect effects are slightly different, except socio-economic status which had the maximum direct as well as the total indirect effect.

Table 4.66 further shows that the majority of the independent variables had their largest indirect effect through socio-economic status (X_6) on the role performance of farm women in this segment.

The results indicating that the effect of socio-economic status, family education and caste on role performance in health care activities of dairy animals revealed that the farm women in higher socio-economic status having high family education but lower in caste are involved more in health care activities. Health care is relatively technical work where more knowledge is required which is generally available in higher socio-economic status families with higher family education. The results in zero order correlation analysis and multiple regression analysis also revealed the same results.

4.7.6 *Role Performance : Milk Products Making*

Table 4.67 indicates that the highest direct effect on the performance of farm women in milk products making was exercised by socio-economic status (X_6), followed by caste (X_3) and age (X_1). Age had negative direct effect whereas socio-economic status and caste had positive direct effects on the role performance of farm women in this segment.

Table 4.67 and Fig 4.13 indicate that the ranks of the variables with respect to the direct effects and the total indirect effects are not the same. Land-holding (X_5), family education status (X_7) and position of respondent (X_2) had maximum total indirect effect, in that order, on the role performance of farm women in this segment. As far as the largest indirect effect through a single variable is concerned, most of the independent variables had their largest indirect effects through socio-economic status and land-holding on the performance of farm women in this segment.

The results obtained through zero order correlation analysis had shown the relationship of socio-economic status, caste and family education with role performance in milk products making. Subsequently, multiple regression analysis could establish the contribution of socio-economic status in role performance. The path analysis concluded the results indicating that socio-economic status is the major variable having direct effect on role performance of farm women in milk products making.

The farm women in families belonging to higher socio-economic status perform the tasks related to this segment at a higher level. This is probably because, generally, such families have more surplus milk for converting it to products like *Ghee*, *Paneer* etc. which requires their women to get involved in the activities of this segment at a level higher than those belonging to lower socio-economic status.

TABLE 4.66: Direct and indirect effects of exogenous variables on the role performance of farm women in the activities/tasks related to health care of dairy animals

(*N = 120*)

Exogenous variables	Direct Effects		Total Indirect Effects		Largest Indirect Effects Through Single Variable	
	Effect	Rank	Effect	Rank	Effect	Variable No.
X_1 Age	0.1674	IV	–0.0770	III	–0.0550	X_3
X_2 Position of respondents	–0.0284	VIII	–0.0122	VII	0.0576	X_1
X_3 Caste	–0.2042	III	0.0991	II	0.1343	X_6
X_4 Family type	–0.0679	VII	0.0168	VI	–0.0695	X_6
X_5 Land-holding	0.0102	X	0.0113	VIII	0.1777	X_6
X_6 Socio-economic status	0.2555	I	0.2539	I	0.1781	X_7
X_7 Family education status	0.2369	II	0.0092	X	0.1917	X_1
X_8 Dairy herd size	0.1334	V	0.0279	IV	0.0569	X_6
X_9 Information sources utilization	0.1086	VI	0.0236	V	0.1248	X_6
X_{10} Economic motivation	0.0115	IX	0.0096	IX	0.0872	X_6

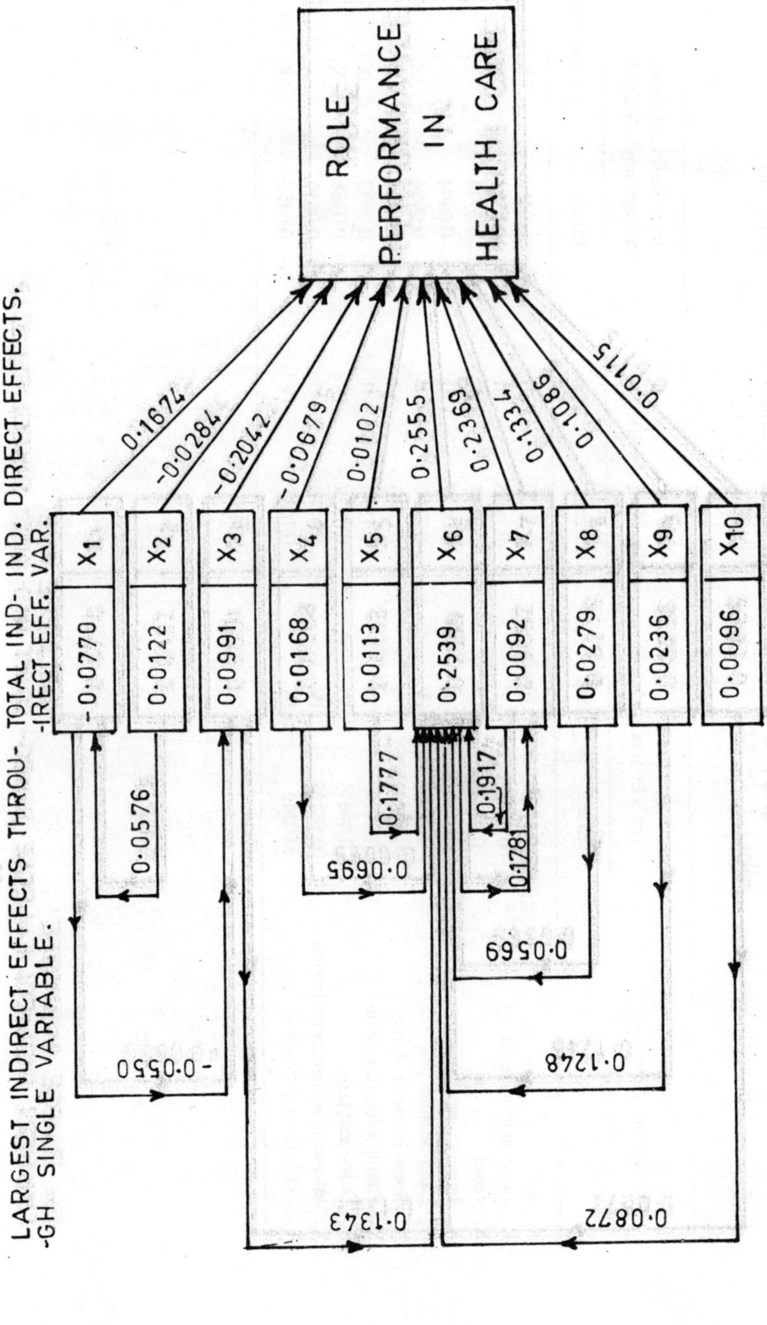

FIG 4.12: Path analysis showing direct and indirect effects of independent variables on role performance in health care

TABLE 4.67: Direct and indirect effects of exogenous variables on the role performance of farm women in the activities/tasks related to milk products making

(*N = 120*)

Exogenous variables	Direct Effects		Total Indirect Effects		Largest Indirect Effects Through Single Variable	
	Effect	Rank	Effect	Rank	Effect	Variable No.
X_1 Age	–0.1514	III	0.0520	VII	0.0460	X_3
X_2 Position of respondents	0.0631	VII	–0.1425	III	–0.0521	X_1
X_3 Caste	0.1711	II	0.0114	X	–0.0733	X_5
X_4 Family type	–0.0116	X	0.0479	VIII	–0.0494	X_5
X_5 Land-holding	–0.1329	V	–0.2200	I	0.0924	X_6
X_6 Socio-economic status	0.1737	I	0.0680	VI	–0.1208	X_5
X_7 Family education status	0.0499	VIII	0.1590	II	0.0997	X_6
X_8 Dairy herd size	0.0947	VI	0.0339	IX	0.0368	X_6
X_9 Information sources utilization	0.1649	IV	0.0770	V	0.0649	X_6
X_{10} Economic motivation	0.0243	IX	0.1087	IV	0.0453	X_6

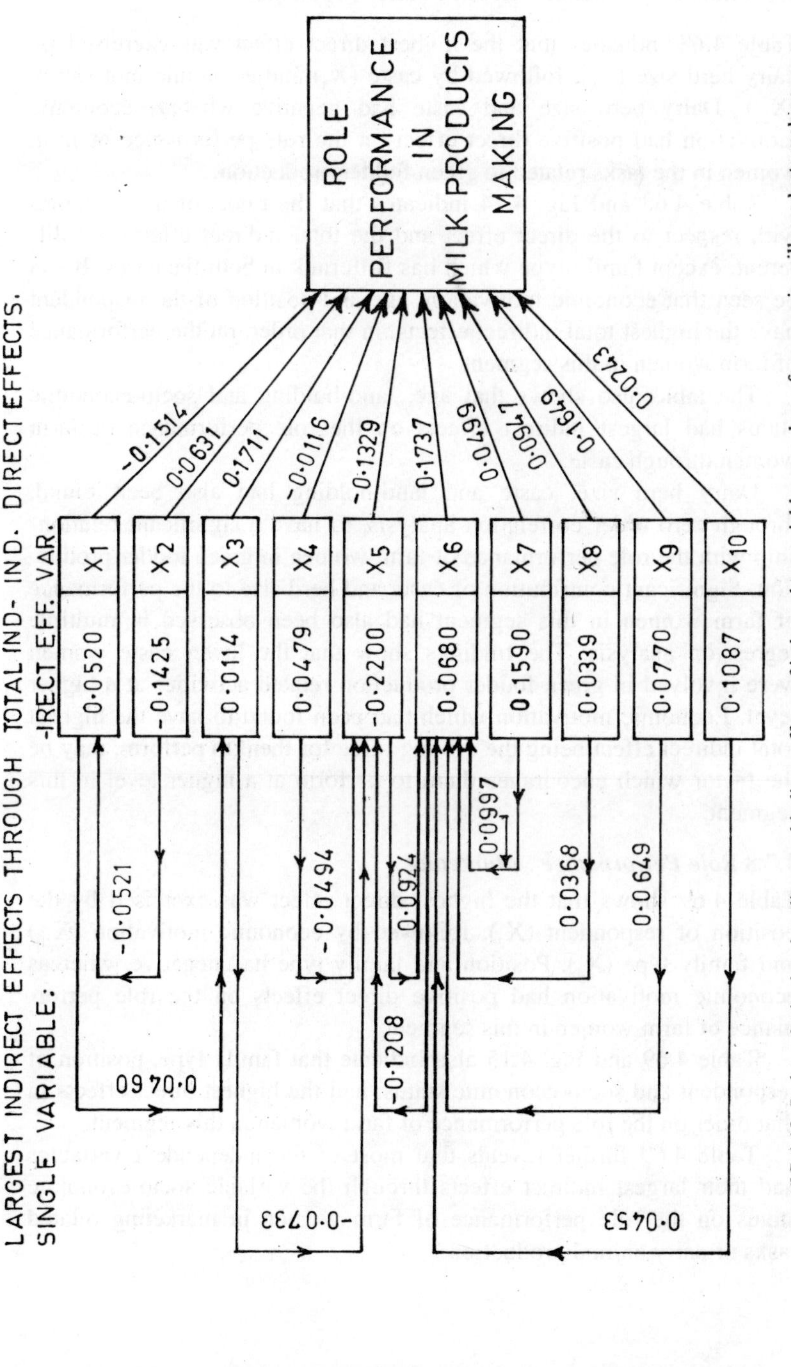

FIG 4.13: Path analysis showing direct and indirect effects of independent variables on role performance in milk products making

4.7.7 *Role Performance : Green Fodder Production*

Table 4.68 indicates that the highest direct effect was exercised by dairy herd size (X_8), followed by caste (X_3) and economic motivation (X_{10}). Dairy herd size and caste had negative whereas economic motivation had positive direct effect on the role performance of farm women in the tasks related to green fodder production.

Table 4.68 and Fig. 4.14 indicates that the ranks of the variables with respect to the direct effect and the total indirect effects are different, except family type which has fifth rank in both the cases. It can be seen that economic motivation, age and position of the respondent have the highest total indirect effects, in that order, on the performance of farm women in this segment.

The table also shows that age, land-holding and socio-economic status had largest indirect effects on the role performance of farm women through caste.

Dairy herd size, caste and land-holding had also been found, through zero order correlation analysis, to have a significant relationship with the role performance of farm women in green fodder production. Significant contribution of caste and herd size to the performance of farm women in this segment had also been observed in multiple regression analysis. The findings show that the lower caste women were involved in green fodder production related activities at a higher level. Economic motivation which had been found to have the highest total indirect effect being the driving force for them to perform, may be the factor which encourages them to perform at a higher level in this segment.

4.7.8 *Role Performance : Marketing*

Table 4.69 shows that the highest direct effect was exercised by the position of respondent (X_2), followed by economic motivation (X_{10}) and family type (X_4). Position and family type had negative, whereas economic motivation had positive direct effects on the role performance of farm women in this segment.

Table 4.69 and Fig. 4.15 also indicate that family type, position of respondent and socio-economic status, had the highest direct effects in that order on the role performance of farm women in this segment.

Table 4.69 further reveals that most of the independent variables had their largest indirect effects through the variable socio-economic status on the role performance of farm women in marketing related tasks of dairy animal production.

TABLE 4.68: Direct and indirect effects of exogenous variables on the role performance of farm women in the activities/tasks related to green fodder production

(*N = 120*)

Exogenous variables	Direct Effects		Total Indirect Effects		Largest Indirect Effects Through Single Variable	
	Effect	Rank	Effect	Rank	Effect	Variable No.
X_1 Age	0.0740	VII	0.1115	II	–0.0581	X_3
X_2 Position of respondents	–0.1474	VI	0.1078	III	0.1244	X_4
X_3 Caste	–0.2159	II	0.0213	IX	–0.0791	X_5
X_4 Family type	–0.1828	V	–0.1002	V	–0.1002	X_2
X_5 Land-holding	–0.1874	IV	0.0569	VII	–0.0911	X_3
X_6 Socio-economic status	–0.0467	VIII	0.1030	IV	–0.1135	X_3
X_7 Family education status	–0.0231	IX	0.0034	X	0.0852	X_6
X_8 Dairy herd size	–0.2330	I	0.0932	VI	0.0930	X_{10}
X_9 Information sources utilization	0.0084	X	–0.0280	VIII	0.0628	X_7
X_{10} Economic motivation	0.2052	III	0.1339	I	0.1056	X_8

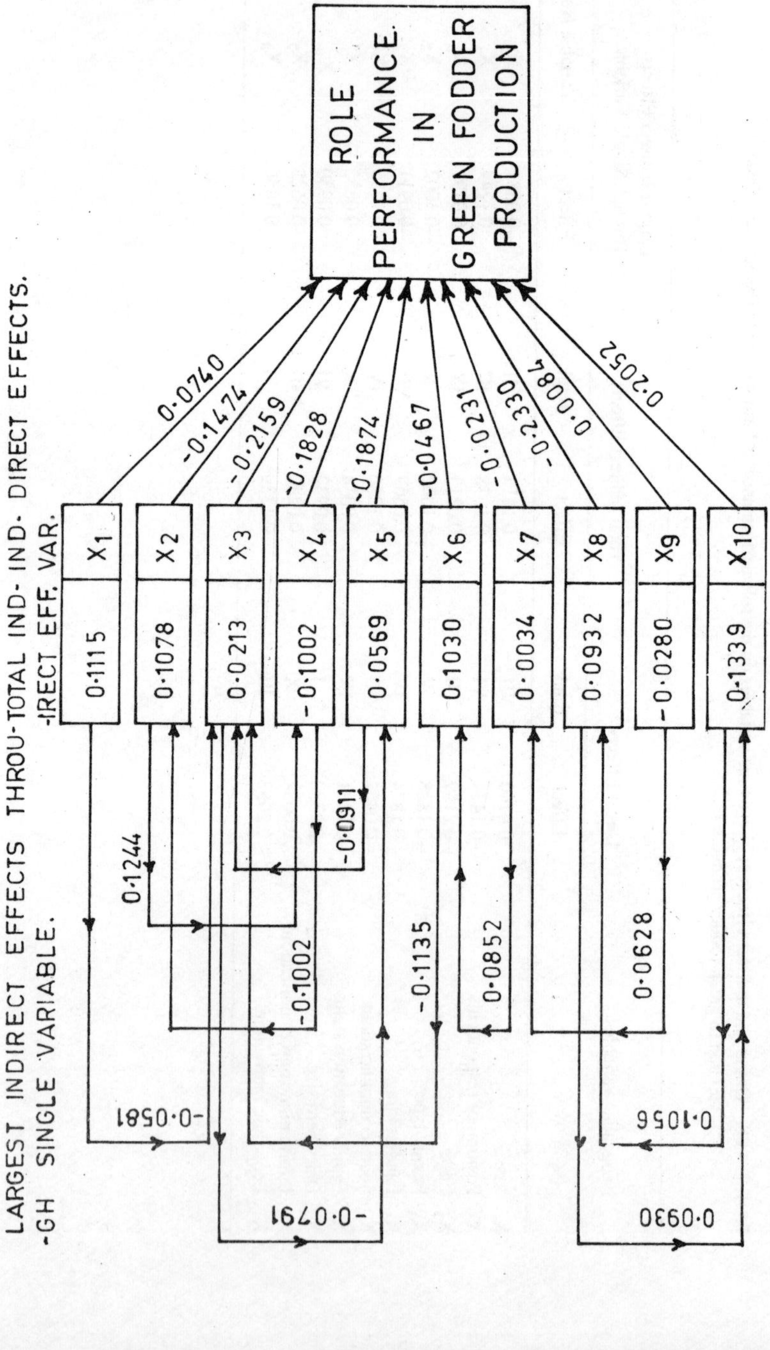

FIG 4.14: Path analysis showing direct and indirect effects of independent variables on role performance in green fodder production

TABLE 4.69: Direct and indirect effects of exogenous variables on the role performance of farm women in the activities/tasks of dairy animal production

(*N* = *120*)

Exogenous variables	Direct Effects		Total Indirect Effects		Largest Indirect Effects Through Single Variable	
	Effect	Rank	Effect	Rank	Effect	Variable No.
X_1 Age	0.1412	IV	0.1026	VII	0.1121	X_2
X_2 Position of respondents	–0.3259	I	–0.1694	II	–0.1140	X_4
X_3 Caste	–0.0493	IX	–0.0892	VII	0.0645	X_6
X_4 Family type	–0.1676	III	–0.1913	I	0.2217	X_2
X_5 Land-holding	0.0758	VIII	0.0064	X	–0.0852	X_6
X_6 Socio-economic status	–0.1227	VI	–0.0175	III	0.0721	X_7
X_7 Family education status	0.0961	VII	0.0566	VIII	0.0920	X_6
X_8 Dairy herd size	0.1310	V	–0.0552	IX	–0.0835	X_{10}
X_9 Information sources utilization	0.0341	X	0.1032	V	0.5990	X_6
X_{10} Economic motivation	0.1844	II	0.1428	IV	0.0594	X_6

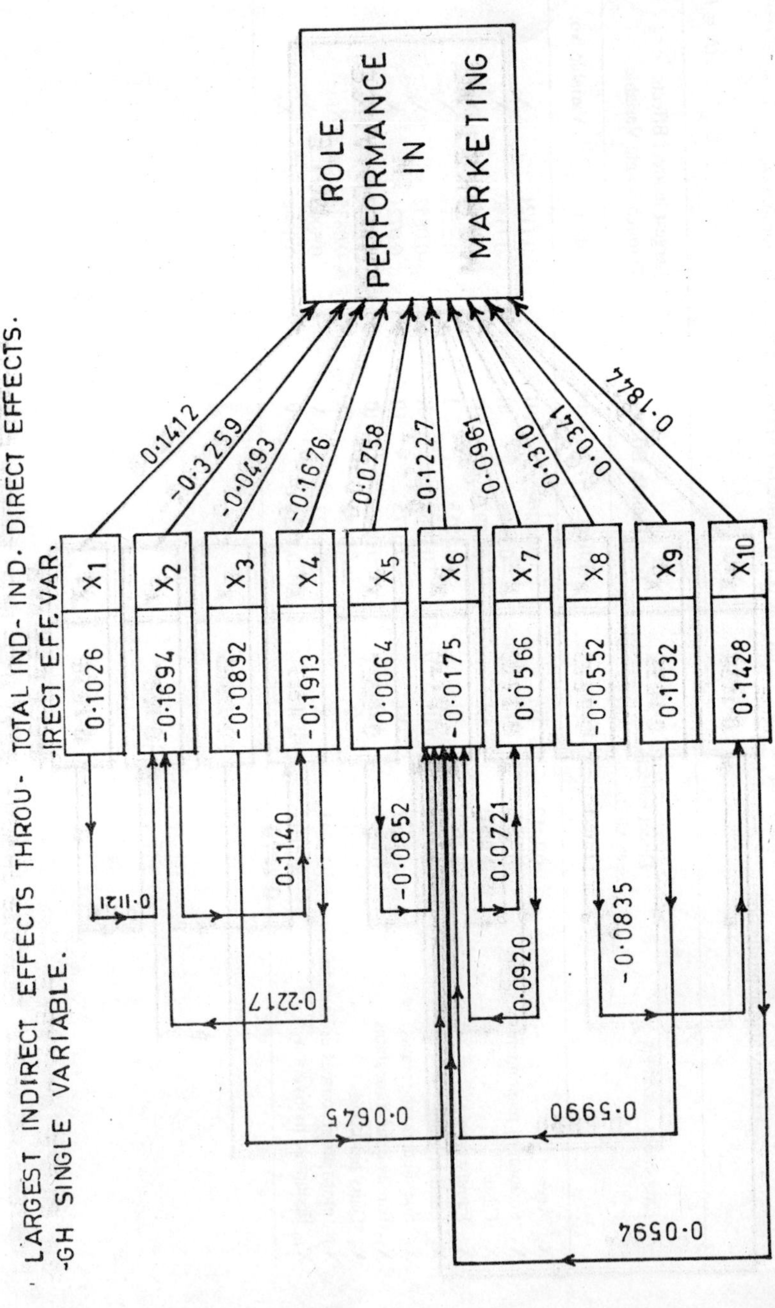

FIG 4.15: Path analysis showing direct and indirect effects of independent variables on role performance in marketing

It may be concluded that all the variables included in the path analysis were important for the productivity of milk and performance of farm women in dairy animal production activities. Three variables viz. socio-economic status, land-holding and dairy herd size were found to be more important for the productivity of milk as these variables contributed substantially, both directly and indirectly to the productivity of the dairy animals.

Five variables viz. socio-economic status, caste, land-holding, dairy herd size and economic motivation were found to be more important for the performance of farm women in dairy animal production activities. These variables contributed substantially, both directly and indirectly towards the performance of farm women in different segments of dairy animal production.

5

SUMMARY AND CONCLUSIONS

This chapter deals with the summary, findings, conclusions, implications and recommendations of the study.

5.1 Summary

Rural women play a significant and crucial role in animal husbandry and allied fields. Studies on women in agriculture conducted in India and other developing countries have indicated that women contribute far more to agricultural production than has generally been acknowledged. The passive and invisible role played by women has resulted in the present state of their general neglect in the developmental processes. They do not have access to training through which they can increase their knowledge and skills to keep pace with the changing rural agricultural scene.

Women participate and contribute in almost all the activities of crop and animal husbandry enterprises. It is difficult to exactly quantify the role of farm women, because of wide variations in the agro-climate conditions, socio-economic and cultural differences under which they perform. The far reaching implications of non-availability of realistic data on their contributions in the growth and development of agriculture, of which animal husbandry is an integral part, provides a compulsive logic for generating empirical data on their role. Because of inadequate and/or partial information, the women are denied their rightful access to development resources and services. This signifies the importance of having a fuller understanding of the role and contributions of farm women so that the extension services may accordingly be atuned to fully integrate them in the developmental processes.

The role of an incumbent in a given social system is the dynamic interplay of the expectations by respective incumbents, his/her own perceptions, his/her own abilities and capabilities governed by personal characteristics and the infrastructural or situational variables. The present study was undertaken to analyse the role of farm women in the seven identified role segments of dairy animal production, the factors governing it and finally its impact on productivity of milch animals. The following specific objectives were framed for the study:

1. To study the role expectations of spouses from farm women in relation to dairy animal production tasks.
2. To study the role of farm women as perceived by themselves in relation to dairy animal production.
3. To study the extent of involvement of farm women and other family members in decision making in relation to productivity of dairy animals.
4. To find out the levels of knowledge and skill of farm women about selected dairy husbandry practices.
5. To study the role performance of farm women in relation to productivity of dairy animals.
6. To study the dynamic interplay of various factors influencing the role of farm women in relation to productivity of dairy animals.
7. To identify the training needs for farm women in relation to dairy animal production.

The available relevant literature was screened and reviewed.

The study was conducted in Bareilly district of Uttar Pradesh. The data were collected through personal interview schedule administered on randomly selected 120 farm women from 12 villages, drawn from four blocks, selected on the basis of purposive stratified random selection technique. The data were collected on 15 independent variables with productivity of milk and role performance as dependent variables. The independent variables studied were: age, position of respondent, caste, family type, land-holding, socio-economic status, family education status, dairy herd size, information sources utilization, economic motivation, decision-making pattern, role expectation, role perception, knowledge and skill. The data on role expectation of spouses from farm women were collected through interview from the respective spouses of the 120 randomly selected farm women. Most of the variables were measured by using the available scales and for measuring skills, the tests were developed for the study.

The relationship between the independent and dependent variables were studied through correlation analysis, while the predictibility of dependent variable was calculated with the help of multiple regression analysis. Path coefficient analysis was done to determine the contribution of each independent variable directly to the dependent variable and indirectly through dominant ones. Other statistical analysis such as analysis of variance, paired 't' test and critical difference test were also used.

The results obtained were as follows:

1. More than half of the respondents were housewives of middle age group (30-40 yrs.). Mothers-in-law and daughters-in-law were almost equal in number.
2. About half of the respondents belonged to lower castes, followed by agriculture caste.
3. About 80 per cent respondents had agriculture as their main family occupation. Most of the families were earning their livelihood from more than one occupation.
4. Only 10 per cent of the respondents were literate. No respondent had formal education above high school level.
5. A very small number of respondents were having *pucca* houses.
6. About 70 per cent of the respondents belonged to joint family and more than 60 per cent families had more than 5 members.
7. Majority of the respondents belonged to marginal farmers' category and possessed 1-2 milch animals.
8. Most of respondents were possessing less than 1 hectare of land and were not cultivating green fodder.
9. Majority of the respondent families owned bicycles. Radio and TV were possessed by very few respondents.
10. Most of the respondents belonged to lower-middle or middle socio-economic strata.
11. The families belonging to the lower socio-economic status had low level of productivity of milk, while upper socio-economic status families had medium to high levels of productivity of milk.
12. Family education status of the respondents was very low.
13. The non-institutionalised interpersonal sources of information were utilized by majority of farm women, particularly within the village. Friends, relatives and neighbours within the village were the main sources of information.
14. Farm women had low level of knowledge in the role segments

of feeding, breeding, management, health care and marketing of dairy animal production.

15. The farm women were found to possess medium to high level of knowledge only in role segment of milk products making.
16. Majority of farm women had low skill levels in the two skills studied viz., clean milk production and preparing and feeding of home made concentrate mixture.
17. Most of the decisions about feeding of dairy animals were taken by housewife alone.
18. Farm women were not involved directly in decision-making of breeding segment of dairy animals. Normally the decisions were taken by spouses only. Very few decisions were taken jointly by the spouse and the housewife.
19. In health care segment, most of the decisions were taken by spouse only.
20. Majority of the decisions in milk products making activities were taken by farm women alone.
21. In role segment of green fodder production, majority of the decisions were taken by spouses only. In some of the farm families, decisions in this segment were taken collectively.
22. In marketing activities, most of the decisions were taken collectively by the family.
23. The farm women were expected by the spouses to perform most of the feeding activities like 'giving feed and fodder to animals', 'preparation of concentrate mixture and giving it to animals', 'mixing green fodder with roughages and giving drinking water to animals'.
24. A high role perception of farm women was observed in feeding activities of the dairy animals.
25. High role performance of farm women was found in feeding the dairy animals. They were performing the activities like 'giving fodder and feed to animals', 'preparing concentrate mixture and giving it to animals' and 'mixing fodder with roughages'.
26. Low role expectation of spouses from farm women, was found in breeding of dairy animals. High role expectation was, however, found in two activities of this segment, viz., 'arranging materials at the time of parturition', and 'taking care of animals at the time of parturition'.
27. The level of role perception of farm women in breeding segment of dairy animal production was found to be very low.

They, however, perceived their role in 'arranging materials at the time of parturition' and 'taking care of animals at the time of parturition' at a high level.

28. Very low performance of farm women was observed in breeding tasks of the dairy animals. Only one activity, viz. 'arranging materials at the time of parturition', was performed by them at a high level.
29. High role expectation of spouses from farm women was found in management of dairy animals.
30. Farm women also perceived and performed their role in management of dairy animals at a high level.
31. Low role expectation of spouses from the farm women was observed in health care activities of dairy animals.
32. Very low performance of farm women, was observed in health care activities, except in two activities viz. 'providing first aid in case of emergency' and 'arranging household medication for sick animals'.
33. Very high role expectation of spouses from farm women, was found in milk products making activities.
34. Very high role perception of farm women was observed in the milk products making segment.
35. Farm women performed the tasks associated with milk products making at a high level except one activity viz., 'straining the milk', where performance of farm women was found to be very low.
36. Very low role expectation of spouses from farm women, perception and performance by farm women was found in the role segments of green fodder production and marketing.
37. In terms of importance the judges assigned the highest ranking to the feeding segment of dairy animal production, while spouses expected the farm women to contribute the most in the segment of milk products making. The farm women's perception and performance were, however, in consonance with the expectations of the spouses.
38. Significant difference was found in role expectation of spouses and role performance of farm women in feeding the dairy animals, among different categories of socio-economic status.
39. Significantly higher role expectation of spouses and higher role performance of farm women were found in the lower socio-economic strata than in higher strata, in feeding of dairy

animals.

40. Significant differences were observed in role expectation of spouses from farm women and role perception of farm women in breeding segment of dairy animal production, among different categories of socio-economic status.
41. Significantly lower expectation of spouses from farm women and significantly lower role perception of farm women were found in the lower socio-economic strata than in the higher strata in breeding segment of dairy animal production.
42. In management of dairy animals, a significant difference was observed in role expectation of spouse, among different categories of socio-economic status.
43. The expectation of spouses from farm women in management tasks of dairy animals, were significantly higher in lower socio-economic strata, than in the higher strata.
44. Significant differences were found in role expectation of spouses from farm women and role perception of farm women in health care activities among different categories of socio-economic status.
45. Significantly low role expectation of spouses and low role perception of farm women were observed in lower socio-economic categories than in the higher strata in health care activities of dairy animals.
46. Significant difference was found in role performance of farm women in role segment milk products making, among different categories of socio-economic status.
47. Significantly lower role performance of farm women was found in milk products making activities in lower than in higher socio-economic strata.
48. Significant difference was observed in role expectation of spouses, role perception and performance of farm women in green fodder production in different categories of socio-economic status.
49. In role segment of green fodder production, significantly higher role expectation of spouses, role perception and performance of farm women were found in lower socio-economic status than in the higher strata.
50. No significant difference was observed in role expectation of spouses, role perception and performance of farm women in activities associated with marketing among different categories

of socio-economic status.

51. Families belonging to higher socio-economic strata and having smaller number of dairy animals had achieved a higher productivity of milk.
52. Productivity of milk of dairy animals was found positively and significantly associated with level of participation of farm women in decision-making pattern in feeding the dairy animals.
53. Daughters and daughters-in-law, farm women belonging to younger age group, lower caste, joint families, lower socio-economic strata and possessing smaller size of family land were found to be performing the feeding tasks of dairy animals at a significantly higher level than others.
54. Economic motivation was found to be positively and significantly associated with the role performance of farm women in feeding the dairy animals.
55. Socio-economic status and family education status, were found to be significantly and positively associated with role performance of farm women in breeding segment of dairy animals.
56. The role performance of farm women in management of dairy animals was found to be negatively and significantly associated with position of respondents and dairy herd size possessed by the family.
57. Socio-economic status was found to be significantly and positively associated with role performance of farm women in health care of dairy animals.
58. Farm women belonging to higher caste, higher socio-economic strata and families having a higher family education status were performing the milk products making tasks at a significantly higher level than others.
59. Caste, land-holding and dairy herd size were found significantly and negatively associated with the role performance of farm women in green fodder production.
60. No significant relationship was found between independent variables studied and the role performance of farm women in marketing segment of dairy animal production.
61. Significant and positive relationship was observed between level of participation of farm women in decision-making in feeding, management and milk products making segments and their role performance in these segments.
62. Role expectations of spouses from farm women in feeding,

management and milk products making were found to be significantly and positively associated with role performance of farm women in these segments.

63. Role perception of farm women in feeding, management and milk products making segments were found to be significantly and positively associated with their role performance in these segments.
64. A positive and significant relationship was found between knowledge level of farm women in feeding and her performance in this segment.
65. Level of skill possessed by farm women in clean milk production was found to be positively and significantly associated with her level of performance in management of dairy animals.
66. The regression model using ten independent variables explained 57.58 per cent variability in productivity of milk, which was found significant ($P < 0.01$).
67. Out of ten, only one variable i.e. socio-economic status, contributed significantly in explaining the variability of productivity of milk.
68. The regression model explained 35.08 per cent variability of role performance of farm women in feeding due to combined effect of all the ten variables studied and was found significant ($P < 0.01$).
69. Caste, socio-economic status and economic motivation contributed significantly in explaining the variability of role performance of farm women in feeding the dairy animals.
70. The independent variables together explained 24.83 per cent variability of role performance of farm women in the breeding segment of dairy animal production and was found statistically significant ($P < 0.05$).
71. Only socio-economic status contributed significantly ($P < 0.05$) in explaining the variability of role performance of farm women in breeding segment of dairy animal production.
72. The ten independent variables included in the study together explained 32.63 per cent variability of role performance of farm women in management of dairy animals and was found significant ($P < 0.01$).
73. Dairy herd size contributed significantly in explaining the variability of role performance of farm women in management of dairy animals.

74. The regression model using ten independent variables, explained 26.05 per cent variability of role performance of farm women in health care of dairy animals and was found significant ($P < 0.01$).
75. Socio-economic status contributed significantly ($P. < 0.05$) in explaining the variability of role performance of farm women in health care activities of dairy animals.
76. The regression analysis explained 26.74 per cent variability of role performance of farm women in milk products making segment and was found significant ($P < 0.01$).
77. Socio-economic status contributed significantly ($P < 0.01$) in explaining the variability of role performance of farm women in milk products making segment.
78. The multiple regression model explained 38.72 per cent variability of role performance of farm women in green fodder production segment and was found significant ($P < 0.01$).
79. Caste and dairy herd size significantly contributed in explaining the variability of role performance in green fodder production segment of dairy animal production.
80. The ten independent variables taken together in the regression model could explain 20.21 per cent variability of role performance of farm women in marketing segment, which was not significant.
81. Socio-economic status and dairy herd size had maximum direct effects on productivity of milk. Maximum number of variables had indirect effects through socio-economic status.
82. Economic motivation, land-holding and family type had maximum direct effects on role performance of farm women in feeding of dairy animals. Maximum number of variables had indirect effects either through land holding or through socio-economic status on role performance of farm women in feeding segment.
83. Caste, age and socio-economic status had maximum direct effects on role performance of farm women in breeding segment of dairy animal production. Most of the other variables had indirect effects through socio-economic status on the role performance in this segment.
84. Dairy herd size, caste and land-holding had maximum direct effects, in that order, while maximum number of variables had indirect effects through caste, land-holding and dairy herd size

on role performance of farm women in management of dairy animals.

85. Socio-economic status, family education status and caste, had maximum direct effects, whereas maximum number of variables also had indirect effects through socio-economic status on the role performance of farm women in health care of dairy animals.
86. Socio-economic status, caste and age had maximum direct effects, in that order, whereas most of the studied independent variables had maximum indirect effects through socio-economic status and land holding on role performance of farm women in milk products making segment.
87. Dairy herd size, caste and economic motivation had maximum direct effects, in that order, while maximum number of variables had indirect effects through caste on the role performance of farm women in the role segment of green fodder production.
88. Position of respondents, economic motivation and family type had maximum direct effects, whereas maximum number of variables had indirect effects through socio-economic status on role performance of farm women in marketing segment.

5.2 Conclusions

1. As younger farm women belonging to lower castes, lower socio-economic strata and having smaller land and live-stock holding, had lower productivity of milk of their dairy animals and knowledge and skill levels, the animal husbandry transfer of technology programmes should be planned accordingly and implemented persuasively through personal contacts with the landless livestock owners, small and marginal farmers and farm women.
2. As in the matters of breeding, health care, green fodder production and marketing activities of dairy animal production, the decisions were taken either by livestock owners alone or collectively by the family. The farm women had very low involvement in decision making in these segments which are important areas in he totality of dairy animal production. There is a need to provide technical know-how and guidance to the farm women to increase their involvement and participation in these areas for an overall improvement in their contribution in the

decision making process.

3. The performance of farm women in the feeding, management and milk products making was found to be high and in consonance with the expectations of the livestock owners and their own perceptions about these segments. Obviously there is no role conflict in these segment between the livestock owners and the farm women. This should encourage the extension policy makers and other extension personnel to plan and formulate the extension programmes accordingly so that farm women could also become an integral part of the programmes and derive suitable benefits from them.
4. Productivity of milk depends largely on the total resources and efforts of the family members in the production activities of which farm women is an important component. Training programmes in dairy animal production catering to the needs of the family as a unit, therefore, need to be developed duly supported by input agencies and credit facilities for increasing the productivity of dairy animals.
5. Knowledge level of the farm women in feeding of dairy animals was found to be positively and significantly correlated with role performance of farm women. It was also found that farm women were involved at a fairly high level in decision-making about feeding and management of dairy animals. With better knowledge about technologies developed in the areas of feeding and management of the dairy animals, the farm women can increase their contribution in augmenting productivity of milk of dairy animals.
6. Farm women from the families belonging to higher socio-economic status, perform their role in breeding, health care and milk products making segments of dairy animal production at a higher level. Participation of farm women from higher socio-economic categories in the extension and training programmes in these segments should, therefore, be ensured by the extension functionaries.
7. In green fodder production and marketing segments, farm women were having least say, the decisions being taken either independently by the livestock owner or collectively by the family members. Proper attention needs to be given by the extension functionaries that the entire rural family is involved in the transfer of technology programmes related to these seg-

ments.

8. In terms of importance of role segments in dairy animal production, judges had assigned top rankings to feeding, management and health care, in that order, whereas according to the perceived importance and performance by the farm women they were giving highest importance to milk products making, feeding and management segments, in that order, of dairy animal production. While the importance given by the farm women to milk products making can be justified to some extent due to their concern as housewife about the nutritional aspect of their family, it can hardly be justified as a livestock producer interested in increasing productivity of her animals by giving due and desired importance to the segments which will give higher production of milk from her dairy animals. Suitable educational programmes, therefore, need to be developed in order to educate them about the relative significance of the segments which contribute more in the augmentation of milk.
9. Socio-economic status had maximum direct effects of productivity of milk of dairy animals and role performance of farm women. The role performance was also affected by land-holding, dairy herd size and caste of the farm women. In order to increase the performance level of farm women in dairy animal production activities and augmenting the productivity of milk, specific extension programmes, including training, need to be developed for the farm women.

5.3 Implications and Recommendations

1. Role of farm women varies from activity to activity. It is the dynamic interplay of several variables such as socio-economic status, land-holding, dairy herd size, caste etc. The study provides the relevant information for identification of the level of the role performed by farm women in a family with relation to dairy animal production. This should help the extension personnel in focusing the extension programmes on the right clientele of the technologies to be trained in specific areas of dairy animal production.
2. In order to achieve higher productivity of milk as well as fodder crops the available technological, financial and manpower resources would need to be fully exploited by educating the

farm families in their proper and optimum utilization.

3. The farm women should be equipped with knowledge and practical skills in feeding, management and milk products making, where they are fully involved. Skill oriented practical training programmes specifically for farm women should, therefore, be developed and organised.
4. Development of suitable and appropriate time efficient technologies to help the farm women in the labour oriented dairy animal production tasks being performed by them will not only increase their efficiency in performing those tasks but will also make more time available to them to perform other tasks more efficiently.

5.4 Suggested Areas of Future Research

1. Production oriented skilled tasks in dairy animal production may be identified and after developing suitable skill tests for them, an experimental study may be taken up to study the individual and collective contribution of the level of skills possessed by livestock owners and farm women in the productivity of milk.
2. A comparative study on the factors affecting the productivity levels of dairy animals may be taken up in a progressive and non-progressive village.
3. A qualitative study of the dairy animal production task, both skilled and unskilled, performed by the livestock owners may be taken up.
4. An experimental study on the qualitative and quantitative impact of training programmes, for livestock owners and farm women on the productivity of milk of dairy animals may be taken up.

BIBLIOGRAPHY

Agarwal, B. (1984). Rural women and the high-yielding variety rice technology. *Economic and Political Weekly* (Review of Agriculture) Vol. 20 (51 & 52), A155-A163.

Agarwal, H. (1987). A study on contribution of farm women in animal husbandry enterprise. M.Sc. Thesis. (Unpub.), I.V.R.I., Izatnagar.

Agarwal, H. (1988). Contribution of farm women in Animal Husbandry. In *Women and Work in Indian Society*, Ed. T.M. Dak, pp. 113-127. Discovery Publishing House, Delhi.

Arya, B.S. (1964). Role of family in decision-making as related to farm operations: A study conducted in Kanjhawala Block, Delhi. M.Sc. Thesis (Unpub.), I.A.R.I., New Delhi.

Arya, S.L. and Madhukar, R.M. (1988). Impact of watershed management on the role of farm women in agriculture in Shivalik foot hills. Paper presented in International Conference on Appropriate Agricultural Technologies for Farm Women, held at ICAR, New Delhi.

Badiger, C.L. (1979). A study on the participation of farm-women in decision-making on farm and home aspects in Dharwad district of Karnataka State. M.Sc. Thesis (Unpub.), University of Agricultural Sciences, Bangalore.

Balasubramanium, R. and Knight, A. John. (1982). Bottlenecks in Modern Dairyman. *Ind. J. Ext. Edn.,* Vol. XXVIII (1&2).

Bareilly District Statistical Bulletin, 1986.

Barua, G.K. (1982). A study of the role of farm-women in agriculture and house making in block Uduguri, Distt. Darang (Assam). M.Sc. Thesis (Unpub.), C.S. Azad Univ. of Agr. & Tech., Kanpur.

Bhangoo, S. (1984). Adoption of improved household practices by farm women as a function of socio-economic factors. *J.Res.* Vol. 23(1), 155-160.

Bhatnagar, O.P. (1971). Role expectation, role performance and training needs of SMS working in J.N. Agriculture University and State Deptt. of Agriculture, M.P., Ph.D. Thesis (Unpub.), A.P.A.U., Hyderabad.

Bhatnagar, D.S. (1985). Selective breeding amount – Recognised Indigenous breed of cattle and buffaloes. Eleventh Dairy Husbandry Officer's Workshop on Transferable Technologies for enhancing milk production, held at N.D.R.I., Karnal.

Bhatt, E.R. (1988). Multiple occupation of farm women: Issues and Implications. A paper presented in International Conference on Appropriate Agricultural Technologies for farm women, held at I.C.A.R., New Delhi.

Bhatt, K.P. (1975). A leading role in agriculture and animal husbandry. *Indian Farming*. Vol. XXV(8), 39-40.

Bhuiyan, A.K.M. (1988). Women's participation in agricultural tasks: An emerging trend in the changing traditional society of Bangladesh. A paper presented in International Conference on Appropriate Agricultural Technologies for farm women, held at I.C.A.R., New Delhi.

Bloom, B.S., Engelhardt, M., Furnșt, E., Hill, W. and Krathwhol, D.R. (1956). Taxonomy of educational objectives. The Cognitive Domain. Longmans Green, New York.

Bond, L., Bible and Coy and Menable, C. (1966). Role consensus and administrative effectiveness. *Rural Sociology,* Vol. 31(1)1, 5-14.

Bose, S. (1985). Caste, tribe and female labour participation. *Social Change,* Vol. 14(2), 15-20.

Chand, S.E. (1967). An investigation into the role expectations and role performance of female members in selected agricultural families. M.Sc. Thesis (Unpub.). Lady Irwin College, University of Delhi, New Delhi.

Chaudhuri, C. (1973). A study on the participation and interest of rural school-going adolescent farm girls in food production and home management. M.Sc. Thesis (Unpub.), I.A.R.I., New Delhi.

Chaudhari, B.N. (1978). Promoting fertilizer use – A study of progressive and non-progressive village. *Fertilizer News,* 89-91.

Chole, R.R. (1978). Role of farm-women in farm decision-making. M.Sc. Thesis (Unpub.), Marathwada Agriculture University, Maharashtra.

Coombs, C.H. (1950). The concepts of reliability and homogeneity. *Educational Psychology Measurement,* Vol. 10, 43-56.

Cravon, R. (1963). The family role in decision-making. *J. Cooperative Ext.,* Vol. 1(1)1, 23-31.

Dairy India (1987). Third Annual Edition. A-25 Priyadarshini Vihar, Delhi.

Dak, T.M., Sharma, H.L. and Jain, R. (1986). Social and institutional framework of female participation in agriculture. *The Ind. J. Social Work*, Vol. XLVII(3), 286-287.

Das, K.K. and Sarkar, D.R. (1970). Economic motivation and adoption of farming practices. *Ind. J.Ext. Edn.,* Vol. 6(1&2), 378-383.

DasGupta, D. and SenGupta, D. (1988). Communitywise involvement of farm women in Indian agriculture. A paper presented in International Conference on Appropriate Agricultural Technologies for farm women, held at ICAR, New Delhi.

Deshpande, W.R. (1962). Some socio-economic aspects of introducing cow dung gas plant. M.Sc. Thesis (Unpub.), I.A.R.I., New Delhi.

Devi, L. (1983). Role expectations and role performance of rural women in farm and home management. Ph.D. Thesis (Unpub.), A.P.A.U., Hyderabad.

Devi, L. and Reddy, V. (1984). Role expectation and role performance of rural women in farming activities. *Ind. J.Ext. Edn.,* Vol. 20(3&4), 27-34.

Davis, M.J. (1957). Decision-making in relation to the performance of household activities in New York State Homes.

Dey, A.K. (1977). A study of decision-making pattern in rural families with respect to selected dairy husbandry practices in Chakka Block of Allahabad Distt of U.P. M.Sc. Thesis (Unpub.), N.D.R.I., Karnal.

Dhillon, J.S. (1982). Some challenges for increasing the productivity of rural milch animals. Proceedings of Tenth North Western Indian Dairy Husbandry Officer's Workshop, held at N.D.R.I. Karnal

Dubey, V.K. (1977). Training lay leaders for dairy development. *Ind. J. Ext. Edn.,* Vol. 38(4), 15-19

Dubey, V.K., Singh, S.B. and Khera, J.K. (1982). Role of rural women in decision-making with respect to animal husbandry practices. *Ind. J. Home Sc.,* Vol. 14(2), 18-21.

Dubey, V.K. and Singh, A. (1975). What cross-bred cattle owner know about scientific dairying. A paper presented at Summer Institute, N.D.R.I., Karnal.

Dubey, V.K. and Singh, S.B. (1976). Technical competence of dairy change agent in ICDP. Unpublished paper presented in 3rd Staff Course, organised at N.D.R.I., Karnal.

Edward Allen (1957). Techniques of Attitude Scale Construction. Appleton Century Crofts Inc., New York.

English, H.D. and English, A.C. (1958). A comprehensive dictionary of psychological and socio-analytical terms. New York, Longman's Green and Company.

Fulzele, R. (1986). Multidimensional analysis of training program of Krishi Vigyan Kendra. Ph.D. Thesis (Unpub.), N.D.R.I., Karnal.

Garg, N. (1966). An investigation into the role expectation and role performance of other key families. M.Sc. Thesis (Unpub.), Lady Irwin College, University of Delhi, New Delhi.

Gatharie, E.R. (1952). The psychology of hearing. Harper and Row, New York.

Gill, S.S. and Batra, P.C. (1967). How to make cattle economics, *Gosamvardhan,* (14).

Gill, S.S. and Singh, P. (1977). Professional knowledge of dairy farmers of Ludhiana District. *Ind. J. .Ext. Edn.,* Vol. 13(3&4), 77-79.

Gite, N.R. (1980). A study of training needs of landless cross-bred cattle owners under Lab to Land program. M.Sc. Thesis (Unpub.), N.D.R.I., Karnal.

Ghosh, S. (1987). Role of village women in decision-making process related to farm economic business. M.Sc. Thesis (Unpub.), Bidhan Chandra Krishi Viswavidyalaya, Mohanpur, Nadia (W.B.)

Ghosh, S. and Adhikary, M.M. (1988). Role of village women in decision-making in different farm economic business. Paper presented in International Conference on Appropriate Agricultural Technologies for Farm Women, held at I.C.A.R., New Delhi.

Goswami, A. (1987). A study of knowledge of the livestock owners about selected animal husbandry practices. M.V.Sc. Thesis (Unpub.), I.V.R.I., Izatnagar.

Govind, S. (1984). Participation of farm women in farm and home activities. M.Sc. Thesis (Unpub.), Tamil Nadu Agricultural University, Coimbatore.

Govind, S. and Subramanyam, V.S. (1988). Participation of women in farm and home activities. Paper presented in International Conference on Appropriate Agricultural Technologies for Farm Women, held at I.C.A.R., New Delhi.

Gupta, P. (1986). A study of communication profile and needs of rural women of Tarai area in Nainital district. M.Sc. thesis (Unpub.), G.B. Pant University of Agriculture & Technology, Pantnagar.

Gupta, S.N., Singh, R.I. and Shukla, S. (1986). Role of women in cattle care. *Dairy Guide,* Vol. 8(11)1, 43-46.

Gulabani, N. (1987). A study of the self esteem of rural women of farm families. M.Sc. Thesis (Unpub.), I.A.R.I., New Delhi.

Gurnami (1985). Transferable Technologies in animal breeding. Eleventh Dairy Husbandry Officer's Workshop on Transferable Technologies for enhancing Milk Production, held at N.D.R.I., Karnal.

India (1984). A reference manual. Publication Division, Ministry of Information & Broadcasting, Govt. of India, New Delhi.

Jackson, M.G. (1978). We need a New Livestock Development Strategy. *Indian Dairyman,* Vol. XXXI(4).

Jain, S.C. and Shukla, A.N. (1988). Prospective avenues of income for farm women through appropriate dairy technology. Paper presented in International Conference on Appropriate Agricultural Technologies for Farm Women, held at I.C.A.R., New Delhi.

Jamal, S. (1984). Suitability of nutrition and home management practices and communication behaviour among farm women. M.Sc. Thesis (Unpub.). G.B. Pant University of Agriculture & Technology, Pantnagar.

Jamal, S., Kunzru, O.N. and Sagar, R.L. (1988). A skill test for farm women – curd making. Paper presented in International Conference on Appropriate Agricultural Technologies for Farm women, held at I.C.A.R., New Delhi.

Jha, P.N. and Singh, B.N. (1965). Utilization of sources of farm information as related to characteristics of farmers. *Ind. J. Ext. Edn.,* Vol. 1, 294-302.

Jhansi Rani, G. (1981). Decision-making patterns among rural families and type of participation by rural women in selected areas of decisions. M.Sc. Thesis, A.P.A.U., Hyderabad.

Kahn, R.L. (1960). Productivity and job satisfaction. *Personnel Psychology,* Vol. 13(3), 275-278.

Kanwar, B.S. and Guleria, J.S. (1988). Role of women in agriculture. Paper presented in International Conference on Appropriate Agricultural Technologies for Farm women, held at I.C.A.R., New Delhi.

Kaur, S. (1966). Participation of farm women in various home and farm activities in changing agricultural situation of Punjab. *J. Research,* Vol. 23(2)1, 346-351.

Kaur, S. (1981). Role of farm women in selected agricultural operations in five villages of Ludhiana district. M.Sc. Thesis (Unpub.), Punjab Agricultural University, Ludhiana.

Kaur, S. (1983). Identification of felt needs of rural women in selected home-making practices. M.Sc. Thesis, Punjab Agricultural University, Ludhiana.

Kaur, S. (1987). *Women in Rural Development: A Case Study*, Mittal Publication, Delhi.

Kherde, R.L. (1971). Role performance and role prediction of the VLW in new strategy of agricultural production. Ph.D. Thesis (Unpub.), I.A.R.I., New Delhi.

King, S.D.M. (1960). *The Integration of Training Organization and Policy: New Development Training*. Helber (Ed.) London. Polytechnic Management Association.

Kolte, N.V. (1967). Study of differential characteristics of farmers having mechanized and non-mechanized farms and problems and factors associated with mechanisation. M.Sc. Thesis (Unpub.), I.A.R.I., New Delhi.

Kumar, R. and Tyagi, K.C. (1985). Role of village Panchayat in dairy development programme. Annual Report, N.D.R.I., Karnal.

Kumari, A., Verma, I., Sinha, I., Srinivasan, V. and Adethi (1988). The churning point - A case study of farm women in dairying in Bihar. Paper presented in International Conference on Appropriate Agricultural Technologies for farm women, held at I.C.A.R., New Delhi.

Kunzru, O.N., Sagar, R.L., Srivastava, N.K. and Singh, P. (1987). Adoption of vaccination against haemorrhagic septicaemia as related to some characteristics of live-stock owners. *Asian J. Dairy Res.,* Vol. 6(4)1, 215-221.

Kunzru, O.N. Sagar, R.L., Srivastava, N.K. and Singh, P. (1987). Constraints in adoption of haemorrhagic septicaemia vaccine. *Ind. J. Ext. Edn.,* Vol. XXIII(3&4), 74-76.

Law, P. (1985). Seventh Five-Year Plan, Rural Development: Its impact on Women. *Social Action,* Vol. 35(1), 29.

Lepcha, T.T. (1987). A study of farm women's contribution to agricultural production in eastern district of Sikkim State. M.Sc. Thesis (Unpub.), Gujarat Agricultural University, Anand.

Li, C.C. (1955). *Population Genetics.* The University of Chicago Press, Chicago.

Mace, C.A. (1950). The analysis of human skills. *Occupational Psychology,* 125.

Malathi, D. (1979). A study of the knowledge and participation of rural women in agricultural operations with respect to Paddy crop and their value orientation in Dharwad district. M.Sc. Thesis (Unpub.), University of Agricultural Sciences, Dharwad Campus, Karnataka.

Mathur, P.N. (1974). Source utilization and rate of spread of information on high yielding varieties of wheat in a farming community. *Ind. J. Ext. Edn.,* Vol. X(1&2), 23-29.

Mehta, P. (1958). A study of communication of agricultural information and the extent of distortion occurring from District to village level workers in selected I.A.D.P. Districts. Ph.D. Thesis (Unpub.). University of Udaipur, Rajasthan.

Moulik, T.K. (1965). A study of predictive nature of some factors of adoption of nitrogenous fertilizers and the influence of source of information on adoption behaviour. Ph.D. Thesis (Unpub.), I.A.R.I., New Delhi.

Nainie, M., Kharde, R.L. and Lokhande, M.R. (1966). The preference of farm women towards food related task. *Ind. J. Ext. Edn.*, Vol. 2(1&2), 41.

Nand, H., Sharma, J.P. and Janardanjee (1988). Contribution of farm women to farm operations and decision-making thereon with special reference to potato. Paper presented in International Conference on Appropriate Agricultural Technologies for Farm women, held at I.C.A.R., New Delhi.

Nataraju, M.S. and Channegowda, M.B. (1985). Comparative study of knowledge and adoption behaviour of small farmers, marginal farmers and agricultural labourers practising dairying. *Dairy Guide,* Vol. 7(12)1, 31-39.

National Perspective Plan for women (1988): A perspective from the women's movement, Report of a debate on 22-23 August, New Delhi.

Ordonez, M. and Goldey, P. (1985). Reading Rural Development. *Communication Bulletin,* Vol. 19, 7.

Paris, T. (1986). Women in rice farming system, Trip report of Thailand and Indonesia (Castilo, G.T.E.D.) 6th to 25th July, 223.

Panwar, S.G. (1979). A study on training needs of members of primary milk producers society in Satara district, Maharashtra. *Ind. J. Ext. Edn.,* 15-19.

Patel, R.K. and Mehta, R.K. (1988). Appropriate livestock production technologies for farm women. Paper presented in International Conference on Appropriate Agricultural Technologies for Farm women, held at I.C.A.R., New Delhi.

Perry, N.C. and Michael, W.B. (1951). The estimation of a phi-coefficient for an entire criterion group from a Phi-coefficient calculated from the use of extreme fails of a normal distribution of criterion scores. *Educational Psychology Measurement,* Vol. 10, 629-639.

Pilot study on the role and awareness of farm women in Jammu Division (1988). Central monitoring and evaluation unit. Agriculture Production Department. Civil Sectt., J & K Govt.

Promila (1981). Adoption of improved selected dairy management practices by farm women in Ludhiana block. M.Sc. Thesis (Unpub.), Punjab Agricultural University, Ludhiana.

Puri, S. (1972). Work roles and decision-making pattern of farm wives and husbands. Ph.D. Thesis (Unpub.), I.A.R.I., New Delhi.

Rajaguru, G.S. and Satapathy, C. (1971). Effectiveness of sources of information in the adoption process of high yielding varieties by the farmers of Puri district in Orissa. *Ind. J.Ext. Edn.,* Vol. VII (3&4), 72-75.

Rao, R. (1976). A study of the influences of selected factors on rural women's participation in agricultural tasks. M.Sc. Thesis (Unpub.), Lady Irwin College, University of Delhi, New Delhi.

Rath, B.B. (1978). A critical analysis of an intensive cattle development project: progress, problems and prospects. Ph.D. Thesis (Unpub,), Hisar Agricultural University, Hisar.

Ray, G.L. (1967). A study of some agricultural and sociological factors related to high and low levels of urbanisation of farmers. Ph.D. Thesis (Unpub.), I.A.R.I., New Delhi.

Reddy, M.S.K. (1962). A study of adoption of improved agricultural practices as a function of some socio-economic factors and sources of information. M.Sc. Thesis (Unpub.), I.A.R.I., New Delhi.

Reddy, B.R. (1983). Role performance and job satisfaction of the VLOs working with Intensive Agricultural Extension Programme (T & V system). M.Sc. Thesis (Unpub.), Andhra Pradesh Agricultural University, Hyderabad.

Rehman, S., Bachu, B.A. Matoo, F.A., Ali, A. and Wani, G.M. (1988). Participation of women in agriculture as perceived by knowledge of scientific poultry husbandry. Paper presented in International Conference on Appropriate Agricultural Technologies for Farm women, held at I.C.A.R., New Delhi.

Report of the National Commission on Agriculture (1976). Ministry of Agr. & Irrigation, Govt. of India, New Delhi, Part VII.

Rogers, E.M. (1962). *Diffusion of Innovations*. Free Press, New York.

Roy, N.K. (1972). Training needs of farmers in relation to High Yielding varieties of paddy. Ph.D. Thesis (Unpub.), I.A.R.I., New Delhi.

Sagar, R.L. (1983). Study of Agro-Economic, Socio-psychological and Extension Communication variates related with the farmer's productivity of major field crops in Harighata Block. Ph.D. Thesis (Unpub.), Bidhan Chandra Krishi Viswavidyalaya, West Bengal.

Sagar, R.L., Kunzru, O.N. and Singh, B. (1988). Management attributes of milk producers: A research report. *Decision,* Vol. 15(2)1, 121-125.

Sagar, R.L., Kunzru, O.N. and Singh, B. (1988). Milk Producer's perception of factors affecting productivity of dairy animals. *Productivity,* Vol. 29(2), 151-156.

Sagar, R.L., Kunzru, O.N. and Singh, B. (1989). Productivity of dairy cows as related to management attributes of milk producers. Proceedings of 76th Session of the Indian Science Congress, Madurai.

Saraswathi, L.S., Padma, A.V. and Perumal, P. (1987). Position of farm women in Tamil Nadu: A report on a general study on farm women in the villages of Keelappavoor (Tirunelveli) and Palur (South Arcot). A research project, sponsored by DANIDA.

Sardana, P.K. (1988). Role of women in agriculture: farm women and dairy supplement one another. *J. Ext. System,* Vol. 4(1), 67-69.

Sargent, S. (1951). Concepts of role and ego in contemporary psychology, *Social Psychology at Cross-road*, ed. by John H., Rohrer and Musafer Sheriff, New York: Harpur and Brothers.

Satapathy, C. (1987). Women in Agriculture: A case study at adopted villages of OUAT. Research project report, Orissa University of Agriculture and Technology, Bhubaneswar.

Sawhney, M.M. (1967). Farm practice adoption and use of information source and media in rural community. *Rural Sociology,* Vol. 32, 310-325.

Seema, B. (1986). Role of women in decision-making process of a farming community in Trivandrum district. M.Sc. Thesis (Unpub.), Kerala Agricultural University, Kerala.

Sharma, A. (1983). Dairy husbandry practices in rural areas. *Rural India J.*, Vol. 46(7-8).

Sharma, D.K. and Singh. T.R. (1970). Participation of women in decision-making process. *Ind. J. Ext. Edn.,* Vol. 6(1&2), 48.

Sharma, S.K. (1971). A study of role expectation, role performance and systemic linkage of the rural institutions in intensive agricultural development programme. Ph.D. Thesis (Unpub.), I.A.R.I., New Delhi.

Sharma, S.K. and Mathuria, D.K. (1985). Female labour participation in rice farming system in Chattisgarh region. *Ind. J. Agri. Econ.*, Vol. 40(3), 235-239.

Sidhu, K.S. (1980). A study of dimensional variation in the knowledge of dairy farmers of ICDP and non-ICDP area of Punjab. M.Sc. Thesis (Unpub.), N.D.R.I., Karnal.

Singh, B.K. (1970). Communication behaviour and social change in relation to high yielding varieties programme in Koshi area. M.Sc. Thesis (Unpub.), I.A.R.I., New Delhi.

Singh, H.S. and Punia, R.K. (1988). Role and status of women in agriculture - prospects and retrospects. Paper presented in International Conference on Appropriate Agricultural Technologies for Farm women, held at I.C.A.R., New Delhi.

Singh, K.N. and Ambastha, C.K. (1975). Communication pattern of farmers: A system analysis. *Ind. J. Ext. Edn.,* Vol. XI(1&2)1, 25-31.

Singh, N.P. (1971). Differential perception of communication source credibility and its utilization pattern among young and adult farmers. M.Sc. Thesis (Unpub.), I.A.R.I., New Delhi.

Singh, N.P. (1981). A test to measure skills of farmers. *Ind. J. Ext. Edn.,* Vol.XVII(1&2), 81-87.

Singh, R. (1971). A comparative study of adoption of high yielding varieties programme by the farmers of television village. M.Sc. Thesis (Unpub.), I.A.R.I., New Delhi.

Singh, T.R. (1968). Participation of rural women in decision-making process related to farm business with special reference to farm families in NES block. M.Sc. Thesis (Unpub.), College of Agriculture, Jabalpur.

Sirohi, S. (1985). Involvement of rural women in farming. *Ind. J. Ext. Edn.,* Vol. 21(3&4), 110.

Sohal, T.S. (1973). Barriers in popularising dairy farming, *Haryana Review,* Vol. VII(7).

Sohal, T.S. and Tyagi, K.C. (1978). Role of knowledge in adoption of dairy innovations. *Ind. J. Ext. Edn.* Vol. 14(3&4), 16-25.

Sridevi, K. (1984). Time management of rural women in home and farm activities. M.Sc. Thesis (Unpub.). Andhra Pradesh Agricultural University, Hyderabad.

Sunita (1977). Decision-making pattern of rural families with special reference to the role of women in selected agricultural practices. M.Sc. Thesis (Unpub.). Punjab Agricultural University, Ludhiana.

Swaminathan, M.S. (1985). Women and rural development. *Mainstream* (Annual number), 113-118.

Tomer, N.S. (1979). Problems and prospects of artificial insemination under field conditions. Paper presented at Seventh North-West Indian Dairy Husbandry Workshop, held at N.D.R.I., Karnal.

Trivedi, G. (1963). Measurement and analysis of socio-economic status of rural families, Kanjhawala Block. Ph.D. Thesis (Unpub,), I.A.R.I., New Delhi.

Tyagi, K.C. (1973). Planning dairy extension programme. *Indian Dairyman,* Vol. XXV(4).

Tyagi, K.S. and Sohal, T.S. (1984). Factors associated with adoption of dairy innovations. *Ind. J. Ext. Edn.,* Vol. XX(3&4), 58-60.

Varma, Shashi Kanta (1992). *Women in Agriculture : A Socio-Economic Analysis*. Concept Publishing Company, New Delhi.

Veena, S., Grover, I. and Munjal, S. (1986). Participation of rural women of Haryana in home, farm and dairy sectors. A report on operational research project, N.D.R.I., Karnal.

Verma, V.K. (1984). Participation of tribal women in decision-making process related to animal husbandry practices. M.Sc. Thesis (Unpub.), Birsa Agricultural University, Ranchi.

Verma, O.S. and Malik, B.S. (1984). Contribution of rural ladies in farm operations. *Home Science,* Vol. 15(2), 57-60.

Weirsma, W. and Jurs, S.G. (1976). Evaluation of Instruction in Individually Guided Education, London. Addison Westley Publishing Co.

Yadav, S. (1982). Role of women in decision-making related to farm business. M.Sc. Thesis (Unpub.), Haryana Agricultural University, Hisar.

Yang, W.Y. (1980). Methods of farm management investigation. F.A.O. Agricultural Development Paper, No. 80, F.A.O., Rome.

INDEX